THE NATURE AND FUTURE OF TOURISM

A Post-COVID-19 Context

THE NATURE AND FUTURE OF TOURISM

A Post-COVID-19 Context

Maximiliano E. Korstanje, PhD
Babu George, PhD

AAP APPLE
ACADEMIC
PRESS

First edition published 2022

Apple Academic Press Inc.
1265 Goldenrod Circle, NE,
Palm Bay, FL 32905 USA

4164 Lakeshore Road, Burlington,
ON, L7L 1A4 Canada

CRC Press
6000 Broken Sound Parkway NW,
Suite 300, Boca Raton, FL 33487-2742 USA

2 Park Square, Milton Park,
Abingdon, Oxon, OX14 4RN UK

Library and Archives Canada Cataloguing in Publication

Title: The nature and future of tourism : a post-COVID-19 context / Maximiliano E. Korstanje, PhD, Babu George, PhD.
Names: Korstanje, Maximiliano E., author. | George, Babu, author.
Description: First edition. | Includes bibliographical references and index.
Identifiers: Canadiana (print) 2021038476X | Canadiana (ebook) 20210384786 | ISBN 9781774637296 (hardcover) |
    ISBN 9781774637340 (softcover) | ISBN 9781003277507 (ebook)
Subjects: LCSH: Tourism. | LCSH: Tourism—Forecasting. | LCSH: COVID-19 Pandemic, 2020-
Classification: LCC G155.A1 K67 2022 | DDC 338.4/791—dc23

Library of Congress Cataloging-in-Publication Data

Names: Korstanje, Maximiliano E., author. | George, Babu, author.
Title: The nature and future of tourism : a post-COVID-19 context / Maximiliano E. Korstanje, PhD, Babu George, PhD.
Description: 1st Edition. | Palm Bay, FL : Apple Academic Press exclusively co-publishes with CRC Press, 2022. |
    Includes bibliographical references and index. | Summary: "This book focuses on the tourism industry in conjunction
    with the impact of COVID-19 from the perspective that it is both negatively impacting the industry while also
    offering it an opportunity to rise from the ashes. The volume offers a new conceptualization and theorization of
    tourism; suggests new research methods, offers parallels with other crises (such as 9-11) to better understand the
    current one, and suggests futurist and innovative strategies. This book offers a wide range of topics that can help
    researchers and practitioners to understand how a pandemic can impact customer satisfaction. The diverse topics
    in The Nature and Future of Tourism: A Post-COVID-19 Context discuss the impact of COVID-19 in the tourism
    industry, application of robotics in the hospitality industry, tourism as a rite of passage, prospects for space tourism,
    how communication technologies can benefit third world tourism, and more. This must-read book not only presents
    sufficient tools for the understanding of how the pandemic can change the industry but is also a thoughtful collection
    of chapters for all those interested about the nature of tourism"-- Provided by publisher.
Identifiers: LCCN 2021058126 (print) | LCCN 2021058127 (ebook) | ISBN 9781774637296 (Hardcover) |
    ISBN 9781774637340 (Paperback) | ISBN 9781003277507 (eBook)
Subjects: LCSH: Tourism. | Hospitality industry. | COVID-19 Pandemic, 2020-
Classification: LCC G155.A1 K65 2022 (print) | LCC G155.A1 (ebook) | DDC 338.4/791--dc23/eng/20211214
LC record available at https://lccn.loc.gov/2021058126
LC ebook record available at https://lccn.loc.gov/2021058127

ISBN: 978-1-77463-729-6 (hbk)
ISBN: 978-1-77463-734-0 (pbk)
ISBN: 978-1-00327-750-7 (ebk)

About the Authors

Maximiliano E. Korstanje

Senior Researcher, Department of Economics, University of Palermo, Argentina; Editor-in-Chief, International Journal of Safety and Security in Tourism

Maximiliano E. Korstanje, PhD, is Editor-in-Chief of the *International Journal of Safety and Security in Tourism* (UP, Argentina) and Editor-in-Chief Emeritus of the *International Journal of Cyber Warfare and Terrorism* (IGI-Global, US). In addition to being a Senior Researcher in the Department of Economics at the University of Palermo, Argentina, he was a Visiting Research Fellow at the School of Sociology and Social Policy, University of Leeds, UK, and the University of La Habana, Cuba. Dr. Korstanje has visited and given seminars at many important universities worldwide. He has also recently been selected to take part in the 2018 Albert Nelson Marquis Lifetime Achievement Award. A great distinction was given to him by Marquis Who's Who in the World. Currently he works as a book series editor of Advances in Hospitality, Tourism, and Service Sectors (IGI Global, US) and Tourism Security and Post-Conflict Destinations (Emerald Group Publishing, UK). E-mail: mkorst@palermo.edu

Babu George, PhD

Professor and Associate Dean, School of Business, Christian Brothers University, Memphis, Tennessee, USA

Babu George, PhD, is a Professor and Associate Dean of the School of Business, Christian Brothers University, a Lasallian Institution, Memphis, Tennessee, US. He has a PhD in Management, a DBA in International Business, and an EdS in Higher Education Innovation and Leadership. He is a graduate of Harvard University's Institute for Educational Management. Previously, he served on a diverse range of academic-administrative roles at various universities, including Fort Hays State University, University of Nevada Las Vegas, Alaska Pacific University, University of Southern Mississippi, among others. In addition, he holds visiting professor designations at more than 15 universities around the world. Since 2001, he has taught a variety of undergraduate, graduate, and doctoral level courses in management, marketing, entrepreneurship, tourism, healthcare, international business, higher education leadership, etc. He has published more than 200 research papers in international scholarly journals.

Contents

Abbreviations

AI	artificial intelligence
ART	Africa Resources Trust
ATT	advanced transport telematics
BBO	buy-build-operate
BDO	build/develop/operate
BOO	build-own-operate
BOT	build/operate/transfer
BTO	build/transfer/operate
CBA	cost-benefit analysis
CBT	community-based tourism
COVID-19	coronavirus disease
CRS	computerized reservation systems
DB	design-build
DBM	design-build-maintain
DBO	design-build-operate
DMG	destination marketing organizations
EUL	developer finance: enhanced use leasing
FARC	revolutionary armed forces of Colombia
GIS	geographic information systems
GPI	global peace index
IAST	International Academy for the Study of Tourism
ICT	information and communication technologies
ITS	intelligent transport systems
JRS	Japanese Rocket Society
LDO	lease/develop/operate
PPP	public-private partnership
SARS	severe acute respiratory syndrome
SC	service contracts
SMEs	small and medium-sized enterprises
UNWTO	World Tourism Organization
VA	veterans affairs

Acknowledgments

To write a book about tourism theory is not easy; however, the task deserved our efforts. The goals of this book aimed at providing a fresh insight that helps scholars to understand the origin and evolution of tourism, as something more complex than a mere industry. At least for authors, tourism is seen as a social institution that traversed different times and cultures. We would thank Apple Academic Press and Sandy for their valuable support, as well as the role of Dr. Regina Schluter and Juana Norrild as authoritative voices in the Latin American tourism research. Over more than 30 years, both have worked hard to keep functioning the leading journal *Estudios y Perspectivas en Turismo [Studies and Perspectives in Tourism],* which illuminated with broader strokes the ways tourism research has been conducted in the region. As a must-read journal, which is closing the doors by these days, EPT offered fertile conceptual and methodological ground for high-quality publications that illuminated the first steps of all of us.

Preface

With the breakout of COVID-19, the tourism industry finds itself forced to navigate and perform into a negative context, where travel is associated with disease, and sometimes death. All the sectors of the industry are impacted, some more than others. Social media sometimes contributes to make the situation even worst, by spreading fear, stress, and distress. Some ethnic groups are sometimes victimized. That said, for some, the breakout of the virus is offering the industry an opportunity to reinvent itself for the best. That said, in reinventing itself, it is important for the industry to be careful in terms of what it wishes for. The impacts of the breakout of COVID-19 have also been on tourism education and related topics such as hospitality, events, leisure, etc. For Seraphin, this pandemic offers an opportunity to review some grounded theories. All in all, it appears that research is highlighting both, the negative and positive impacts of the virus on the tourism industry. This provides another opportunity to recall the Janus-faced character of the industry, but also the ambidextrous perception of academics regarding the virus. As a result, to deal with this virus, tourism stakeholders, and more specifically, decision-makers, need to adopt a Janusian thinking approach. Janusian-thinking, and ambidextrous management approaches have both been associated with innovation, sustainability, and performance improvement.

This book is in line with current research in the area of tourism (and related topics) and COVID-19, as it discusses COVID-19 from the perspective that it is both impacting the industry negatively, while offering it an opportunity to rise from the ashes. In a nutshell, the book has adopted a Janusian way of thinking. To do so, the book offers a new conceptualization and theorization of tourism; suggests new research methods; offers parallels with other crisis to better understand the current one; and futurist and innovative strategies are suggested. These futurist thoughts are extremely important. The purpose is not about forecasting the future of the tourism industry as forecasting in tourism is impossible as the industry is by nature variable. That said, despite the limitations of forecasting, there is a growing need for such exercises to assess the likelihood of changes. It would be interesting to investigate the impacts of COVID-19 on customer satisfaction. More specifically, how has customer satisfaction for a particular service changed under the influence of the virus? Resort mini-clubs or kids' clubs, a service for children when on holidays in resorts, could be taken as an example.

Studying resort mini-clubs is about studying social networks. This type of research involves examining the social system, environment, and context. Indeed, mini-clubs are part of a network of interconnected products and services offered by resorts to their customers. There are two main types of networks, namely the centralized networks where all the nodes are highly dependent on the main node, meaning that this node influence how the entire organization work, and peripheral networks, with isolated nodes interacting between themselves while still interacting with the main node. Mini-clubs are nodes within a peripheral network. Mini-clubs are connecting services and departments such as catering; room services; entertainment/animation; but each of these service/department is forming its own cluster or subgroup. As with any service, the role of a mini-club within the network of services, is to satisfy customers' wants and needs. The latter is the result of positive emotions, which subsequently lead to repeat re-patronage behaviors and positive (e)word of mouth. These positive emotions arise as an evaluation of customers regarding whether or not (a) the products and services promised were provided; (b) interaction with others, staff included; (c) the environment in which the products and services were provided. Equally important, Drewery and McCarville, highlighted the fact that: 'leisure services may be particularly emotional in nature;' 'leisure settings and the nature of the activities themselves all build emotional buy-in on the part of participants.' This book offers a wide range of topics that can help researchers and practitioners to understand how a pandemic can impact on customer satisfaction. More importantly, the book presents sufficient tools for the understanding of how the pandemic can change the industry.

—*Hugues Seraphin, PhD*
University of Winchester, UK

CHAPTER 1

A New Conceptualization of Leisure

ABSTRACT

Over years, sociology, and anthropology have systematically overlooked the important role played by leisure to regulate the different conflicts and cleavages happened in society during the working days. A young Norbert Elias acknowledged the importance of leisure not only revitalizing social frustration but socializing the mainstream cultural values of society. This chapter interrogates furtherly on the different academic waves which have focused on the problem of modern leisure. From different angles, each one has limitations and notable advances in the field. We start the chapter with a historical and conceptual insight on ancient leisure. The efforts devoted to this chapter aims to validate that tourism should be understood as a form of leisure. In the successive sections, we dissect critically the main theories in leisure research laying the foundations to a new agenda for the years to come.

1.1 INTRODUCTION

The etymology of the term *leisure* comes from *leisir (French)* which means *free time.* Although Ancient Greeks used the term *Scholé (σχολή)* to signal to leisure time, the connotation was not the same-nor the meaning over the years-. Scholé was originally reserved to a free time dedicated to contemplation, not necessarily to entertainment. For some reason very difficult to precise here, the word gradually derives in Latin from Scholé to Schola where we obtain finally the words *scholar* and *school.* This happened because education and escapement were inextricably intertwined in the Greek culture (Balme, 1984). During the era of the Roman Empire, schola mutated to *otium,* which was a term reserved only to the leisure time as well as the needs of play or evasion (Toner, 2013). In some respect, *otium* (leisure) was symbolically opposed to *negotium,* a term limited or exclusively used to the daily businesses (Kyle, 2012). As Korstanje (2009) puts it, no matter the time or the culture, the leisure accompanied the history of sedentary societies-even ancient empires-from their inception. Over the centuries, leisure played a crucial role not only because it keeps the society united, but also because the mainstream cultural values of society were venerated and accepted. The leisure and event management were key factors for Roman Emperors and authorities to keep their legitimacy over society. In public baths, celebrations, travels, banquets or the well-known gladiator combats, the cultural values that made from Rome an empire were culturally replicated-if not reaffirmed. Anthropologically speaking, any lay-citizen, regardless of his or her status or wealth was proud to be roman, to belong to a privileged and educated civilization while differentiating her or himself from barbarism (Korstanje, 2009).

In this backdrop, the present chapter discusses critically the main theories, schools, and thinkers who focused their attention on the connection between leisure and society. From Norbert Elias to Erich Weber-only to name a few-without mentioning Frederic Munné, who have systematically conceptualized the leisure as a space of evasion where the day-to-day psychological frustrations are regulated and sanitized. Scholars who paid their attention to leisure were scornfully undermined by other academic waves simply because the idea of free time not only was erroneously associated to emotionality but also because it was negatively thought as the opposite to the western rationality. Positive sociology and anthropology overlooked historically the importance of leisure in the structuration of the social bondage as well as modern politics (Wynne, 2002; Rojek, 2013; Kelly, 2019). What is more important, leisure has changed its meaning according to culture and time (Kelly, 2012). To fill this gap, the current chapter, which is vital to follow the rest chapters in this book, unearths the seminal and empirical contributions of authors who have devoted their life to expand the current understanding of leisure. The first introductory section explains the main contributions and limitations of Frederic Munné, a well-known expert who has devoted his efforts in the study of leisure as well as Weberian tradition, a vital conceptual corpus that illuminated the paths of scholars since the inception of the discipline. Even if Munnè's texts have been never translated into English, his legacy is of vital importance to develop the present chapter. He has reviewed not only the history of leisure but also launched to find a catch-all definition of the phenomenon.

Secondly, we explore the anthropological nature of leisure, discussing the main theories and authoritative voices such as Weber, Dumazedier, Rojek, Elias, and Dunning. Each author keeps his own definition of leisure, but all they shared a similarly-minded thesis. Leisure plays a leading role in enhancing social reciprocity, keeping people together under the same rules and social norms. The third section re-conceptualizes the commonalities and differences between tourism and leisure. While some authors toy with the belief that tourism and leisure are inextricably intertwined, others go in the opposite direction holding that tourism should be framed as a sub-field of leisure studies. Lastly, we review in-depth the intersection of leisure in a hyper-globalized world, without mentioning the risks and problems of consumerism today. In the final section, some remarks about the crisis, as well as the next challenges to face in leisure studies, are carefully discussed.

1.2   FREDERIC MUNNÉ (AN INTRODUCTORY READING)

One of the risks in approaching an all-encompassing definition of leisure seems to be associated to the conceptual framework, one adopts in such a task without mentioning the knowledge fragmentation that divided academicians even to date (Stebbins, 2011; Henderson, 2011). This point led Frederic Munné to review a more than interesting and landmark book, where he discusses the ebbs and flows of each school which has pinned out leisure as their main object of study. Unfortunately, the book *Psicosociologia del Tiempo libre (psycho-sociology of leisure),* a seminal project authored by Munnè was originally written in Spanish, and never translated to English. Because of this, our efforts are intended to present his argumentation as clear as possible alternating their texts in dialog with other authors. Frederic Munné is a cultural theorist-born in Barcelona Spain-who has extensively focused on the intersection of leisure and free time in the societal order. Based on the premise that leisure should be defined as a political (intermixed) act that leads man to a philosophical contemplation, he argued convincingly that any agent debates between two contrasting tendencies, the needs of adopting rules (enhancing its own security) and the liberty which is proper of human nature (moving towards an unknown place). Combining the two tendencies, leisure engages with a sentiment of compensation that starts a much deeper process of revitalization. As

Munnè in his text reminds, four academic schools widely systematically approached leisure: The German School, the Soviet school, the French school, and the American school. Each one maintains their commonalities but conserving serious discrepancies with the rest. The German school is mainly based on a critical approach centering on the advances of anthropology whereas the soviet school paid attention to the preliminary remarks of Karl Marx. Contrariwise, French, and American schools adopted the functionalist paradigm that helps understanding leisure as foundational institution mainly oriented to alleviate the social frustrations that normally place the harmony of society in jeopardy (Munné and Codina, 2002). In a landmark book, entitled *Psicosociologia del Tiempo libre (psychosociology of leisure),* Munné (1980) goes farther and presents an erudite review of the theories revolving around leisure. In his viewpoint, men move in quest of their freedom but paradoxically they are never free. They are constrained to the societal rules that precede them, or so to speak, the preceding rules that were there earlier they are born. Neither the absolute freedom nor the absolute duty exists in the community. Echoing David Riesman-in his book the Lonely Crowd-, Munné holds that societies gradually evolved to an inner-directed man towards an Other-character typology where the self is previously determined by external forces, such as journalism, mass media, and the peer's esteem. The quest for discoveries and visiting exotic places are certainly characteristic of societies based on a society directed to the Otherness. The inner-directed personality impels the individualism and self-contemplation (i.e., Puritanism) while in the other-directed society the self is determined by the opinion of others (capitalism). Travels, explorations, and discoveries are possible in Other-directed societies. In view of this, leisure encompasses the three basic components, originally described by Erich Weber, *regeneration, compensation, and idealization.* While regeneration refers to the physical needs to relax and escapement, compensation only is possible when the idealized goals are reached. The idealization, complementarily, follows a spirit of contemplation that takes place when the person reaches a state of self-fulfillment. Having said this, leisure combines functions of preservation and revitalization with play. In modern capitalism, leisure takes a different meaning associated with consumption, a right or simply a time/space for relaxing purposes. Munné coins the term *bourgeois leisure* to denote the rise of a new practice where the spiritual contemplation is placed in the dust of oblivion. In fact, the philosophical free time linked to the urgency to understand the world sets the pace to a new leisure form psychologically motivated by the consumption and evasion. This externally-designed form of recreation-far from emancipating or educating-oppresses the lay-citizen as never before. Of course, the bourgeois leisure was the target of classic and neo-Marxist scholars who exerted a radical and caustic view of leisure as an instrument of indoctrination, discipline, and control orchestrated by ruling elite to keep the power in its hands. There is a type of new leisure widely commercialized, commoditized where money-exchange and profitable relations prevail. It is important to mention here that the modern man pays to experience a moment of leisure. The worker spends (if not waste) his money (surplus) while capital-owners offer an infrastructure of consumption orchestrated to build a road towards the pseudo-happiness (alienation). At the bottom, in Munné account, modern tourism represents a commoditized version of leisure. Last but not least, he makes finally a bridge between Erich Weber and Dumanzedier, alerting on the risks and problems of modern leisure. As we already reviewed in the introductory chapter-in MacCannell's works-leisure situates as the antithesis of labor, but-at the same time-reproducing the oppressive institutions that cement the authority of capitalism over the workforce.

1.3   DISCUSSING THE NATURE OF LEISURE

In 1899, the Norwegian American economist Thorstein Veblen published a trailblazing treatise under the title of *the theory of the leisure class*. In his work, he argues convincingly that society is structured according to two different-if not opposing-strata: *productive and leisure classes*. On one hand, he is interested in developing an all-encompassing theory that explains human evolution since the feudal period to industrialism. In so doing, he finds that the ruling elite often manipulates and disposes of the means of production situating itself as an unproductive class. The figure of pecuniary emulation serves to transfer status to the privileged group where high-status members reserve the right to consume-without working. Instead, the working class is doomed to be exploited by the upper classes. In this way, the leisure class is represented by members who look to retain high-status positions while controlling the manufacturing surplus. Examples of the professions, which take part in leisure class, are warriors (soldiers), Priests, scholars, and politicians (Veblen, 2017). Of course,

Veblen-though widely criticized by his literary style-was observant of the American lifestyle and its propensity to mass-consumption. Let's explain to readers that in his development, the notion of leisure has a pejorative connotation that today is not shared with other academicians.

One of the most authoritative voices in the leisure studies was German scholar Erich Weber. He interrogated furtherly the bourgeois leisure stressing on the needs of education which re-channels the personal drives to constructive-not destructive-situation. Per Weber, the organization of free time depends on the possibilities of the self to convert his vital space in a rewarding experience. Hence leisure should be understood as a human platform towards self-achievement (Weber, 1969). Weber's argumentation coincides partly with Joffre Dumazedier, who called the attention that sociology needs to understand leisure as a vital force of societal structuration. Probably influenced by the positivism of Durkheim's legacy, French sociology relegated the leisure to recycle bin of the intellectual tradition. The capitalist industrialism has been notably expanded, in which case, the lay-worker seems to be subject to countless deprivations. Meanwhile, leisure has been increased to guide the new free time of workers. He starts from the premise that the modern man is motivated to dispose of his life as he wants. To wit, leisure follows some liberating trend that gradually leads to hedonism and self-gratification (Dumazedier, 1967, 1999). Norbert Elias and Erich Dunning (2008) stress that leisure is inextricably interlinked to the evolution of Western capitalism. As a controlled version of wars, leisure activates the cooperation and loyalty as vital forces of in-group formation but at the same time, it ignites a state of rivalry or confrontation with other groups. Centering their analysis on hooliganism and sports, Elias, and Dunning explain that the modern capitalism represses the natural human emotions, pushing them to the private sphere, and in so doing the self is deprived from its inner-world. The archetype of masculinity and western rationality appeal to emotions as an irrational behavior a pathology which should be dully regulated or repressed. Leisure and games re-draw the necessary landscapes for these emotions surface. Emotions not only are controlled but also commoditized through the entertainment industry. To put the same in bluntly, leisure serves as a dream-like mechanism created to reverse the negative effects of rationality, substituting it by happiness, sadness, fear or even rage. One of the methodological problems of modern sociology rests on the dissociation between work and leisure. For this hegemonic paradigm, free time and leisure are inevitably defined as the lack of work. Starting from the thesis that the archetype of labor equals rationality, one might speculate that leisure leads to laziness and mediocrity. It is important not to lose the sight of the fact that scholars, who embraced this tradition, strongly believe that leisure revitalizes the frustrations of the working day but never escapes successfully to the logic of work.

In contrast to the positive tradition widely cultivated in France, in the United Kingdom, a new academic wave headed by Norbert Elias is turning the attention to leisure as an emergent social phenomenon. For Elias and Dunning – like Dumazedier or Weber – the phenomenon has three clear cut elements: sociability, mobility, and imagination. While sociability signals to a force disposed to foster the enjoyment of players, the competition, and rivalries among gamers – sooner or later – emerge. When this occurs, the gamers or participants are exposed to a risky situation. Under some conditions, the rules of games are simply broken when participants adopt bad behavior. The civilizatory process engages with leisure to expand alternating moments of pleasure and joy with risk and violence. As a disciplined form of war, leisure seems to be inextricably intertwined to *the quest of excitement*. As authors adhere, the evolution of Western capitalism – in the 18th and 19th centuries – corresponded with the evolution of sports. For Dumazedier (1967), leisure excludes social duties or obligations as friendship, or family because it ushers the citizen into the quest of pleasure maximization. He develops the notion of leisure according to the interplay of relaxation, entertainment, and personal development.

In consonance with this, Karl Spracklen (2011) reminds that leisure is torn in a paradoxical situation. On one hand, it associates with the freedom of life, while – on another – it is encapsulated on the rule of the city, which means the authority of the King. Over the years, the theory has evolved towards three different *allegories: leisure as free choice, leisure as socially constrained, and leisure as a completely free choice.* Different scholars have approached leisure form their respective ideological frameworks. Liberals studied the history of leisure as a free or democratic practice, while structuralism emphasized on the role played by structure in leisure practices. Postmodern writers alluded to leisure as a complex net of relations in the constellations of power. Like Habermas who conceives the world according to two specific rationalities, Spracklen recognizes that the dichotomies revolving around liberty and constraint are solved when we apply the Habermassian model. The first is oriented to a communicational process, a dialog with freedom and democracy whereas the second signals to instrumentality which is based on the current economic means of production. Having said this, leisure-inevitably-follows the dynamic of capitalism. He goes on to write:

> *"Leisure as a meaningful, theoretical, framing concept; and critical studies of leisure are a worthwhile intellectual and peda- gogical activity... Indeed, leisure is the part of our lives where the tension between freedom and constraint-agency and structure, resistance, and control-is most visible, so understanding leisure is even more essential as the world and its societies become increas- ingly commodified and ordered. Following Habermas, examining leisure actions can help us understand the conflicting pressures of instrumental control and individual will- and in doing this, critical studies of leisure can and should continue to play a central role in understanding society"*   *—(Spracklen, 2011, p. 6).*

In view of this, Spracklen introduces his readers into an interesting epistemological debate respecting how to study leisure practices. Beyond the hegemony of critical studies which punctuated on the negative func- tion of leisure, scholars should pay attention to history as the mirror that reflects our habits and customs. The way leisure was practiced in Rome varies respecting to our customs and vice-versa. For this reason, it is vital to situate the nature of leisure within the history and the philosophy of leisure. In some respect, leisure has received different connotation along with the time. Ancient philosophers distinguished the good (pleasure) from the bad leisure (excess) insofar as the medieval thinkers-most probably influenced by Christianity-warned energetically on the ethical problems of leisure. In the modernity, not only the liberal thinking has certainly revived but also defines leisure from the lens of free choice and democracy. Citizens-regardless the class, gender or ethnicity-has the right to cultivate leisure practices to show he or she belongs to a liberal society, a community where the individual decision prevails.

As the previous argument is given, Chris Rojek (2005a) brings some philosophical reflections on the autonomy of the agency in what he dubs as *leisure choice*. In the threshold of time, scholars have debated to what extent the agent is free to practice or not leisure, or even if he is determined by the social structure. As Rojek notes, two significant conceptualizations should be brought to the foreground: *embodiment and emplacement.*

Embodiment refers to the fact that leisure actors move as embodied in their biological and biographical body, where the agent is all days dying, whereas the emplacement is determined by the local environment where leisure agents grow and move. He understands the *action approach* as an attempt to introduce relativism in leisure studies has serious problems in understanding the nature and dynamic of politics. In sum, he offers a conceptual model where leisure occupies a central position as an accumulator of social capital to manage ethical obligations, expectation, and knowledge constituting the *non-pecuniary elements* of society. To put the same in other terms, leisure helps in enhancing not only a deteriorated sense of us-living-together but human solidarity.

> *"Social capital increases the wealth of the community by building mutual, reciprocal, obligations that enhance the quality of life. Because leisure is one of the principal institutions which this time allocation is realized it follows that leisure is essential in enhancing social capital. Play is a mechanism for discharging cognitive and motor energies. It also possesses a strong capacity, because playing together involves mutual recognition and support. Of course, play, and leisure sometimes involve mutual aggression and conflict"*
> *(Rojek, 2005a, p. 16).*

The above-cited excerpt echoes original concerns emanated in Elias and Dunning confronting directly with the dominant discourse in leisure studies that used to reign in the Academia. Leisure-far from being a harmonic and stable action-is sometimes conditioned by rivalry and conflict. Among the political aspect of leisure Rojek enumerates three clear forms: *empowerment, distributive justice, and social inclusion.* The empowerment marks the motivation and knowledge disposed in order for the leisure practice to take shape. Distribute justice signals to the allocation of economic and capital resources to grant access of lower classes to leisure practices. However, social inclusion connotes the necessary extension of citizen rights of those citizens who have been historically relegated or marginalized from leisure consumption. In a seminal book published in 2005 (*Leisure theory: principles and practices)* Rojek refines his original thesis. We must depart from the belief that the post-Hobbesian scholars have misjudged the nature of leisure, as a zero-sum institution where the leisure practices engage with conflict. This means that my leisure practices are opposed to my neighbor's preferences and vice-versa. We must think leisure as a platform to create active citizens who struggle to enhance social reciprocity.

Underpinned into the informational society, modern men are influenced by several flows of information increasing their opportunities to be familiar with their rights as active citizens, as Rojek suggests.

> *"Leisure was never free time. But a good deal of it is now absorbed in risk aversion, from cutting down on drinking and stopping smoking, to monitoring diet and checking on terrorist alerts when we visit foreign countries, practicing safe sex and keeping informed of the environmental hazard posed by carbon gas emissions. Now, it is manifestly not the case that active citizens virtuously devote all of their leisure to being observant about risk. But it is the case that when individuals pursue leisure practices that are known to carry risks to themselves and others, ethical considerations are now automatically invoked"*   *(Rojek, 2005b, p. 4).*

This leads us to think that information appears to be a key factor that not only molds leisure practices but ethics. What today can be considered a desirable leisure activity, i.e., smoking or alcohol drinking-may be in the future risky behavior. Rojek holds a more than interesting point defining leisure as an important or-so to speak-a foundational institution originally designed to regulate the internal conflict through *the combination of identity and representation.* In accordance with other scholars such as Weber, Dumazedier or even Elias, he reportedly assumes that leisure facilitates the social background to keep the society united in the threshold of time. As he formally cites:

> *"The institution of leisure performs an important and increasingly prominent role in managing these issues by acting as the basis for identity formation, the representation of solidarity, the achievement of control and challenging unsatisfactory resource allocation and civic regulation through resistance. The changes of the 1960s and 1970s transformed the logic of economic distribution, status allocation and civic regulation that governed our relations to the employment market, to the state and each other. For one thing, capital, and labor resources ceased to be concentrated in industrial and factory production and switched to communication and information. At the*

same time, industrial production and assembly shifted to low labor-cost economies in the developing world" (p. 8).

To cut the long story short, the importance of leisure in conceiving a more collective society is not given without the rise and expansion of globalization. This process paradoxically framed more politicized agents who often confront, digest, and internalize the dominant narratives about how leisure should be practiced. What is more important-echoing Rojek-, leisure practices are freer and more open because society distributes and recognizes the new rights of their citizens. In any case, these liberties engender inherited risks which will be dully addressed in the sections to come. But can leisure escape from the conception of politics? is leisure a modern invention proper of Euro-centrism or a real millenarian institution which can be traced back in history?

1.4   LEISURE STUDIES AND RACISM

Despite the abundant information and the multiplication of recent publications that debate the intersection of power and leisure, less attention was given to the structuration of racism and its evolution through leisure practices. As Mary Louise Pratt eloquently noted, Euro-centrism has historically consisted in marking the "non-western Other" while the privileged group is systematically unmarked. Travels, and the cultivation of arts, as well as the scientific project, are seen as the peak of the iceberg in the racial hierarchy where white Europeans are placed on the top. At the same time, the idea of civilization cannot be grasped without the figure of the primitive mind, blackness cannot be understood without *whiteness*. In this vein, the literature says little respecting to the role of whiteness as the precondition for modern leisure. To put the same in other terms, in what way the notion of race still remains as a key factor that divides practices and identities in the leisure fields. In a more than a pungent book, which entitles *Whiteness and leisure,* Karl Spracklen (2013) analyzes the problem of race as an ideological category which cements the authority or control of a so-called superior group over others, a point that is interlinked to the material asymmetries proper of capitalism. He starts from the premise that it is necessary to reformulate new alternative answers to the problem of race and inequalities, as well as the influence of sports and leisure in the process.

"...The way in which sport is used to construct whiteness, and the way in which whiteness shapes sports, is an important unanswered theme in the sociological critiques of sport, 'race' and racism. This gap in the theory and the empirical research is also seen when the focus of the myth-making-sport-is expanded to include the entire range of people's leisure lives, leisure activities and leisure spaces" **(Spracklen, 2013, p. 4).**

Following this, as he puts it, the belief in whiteness is implicitly encapsulated in our cultural organization. Nonetheless, whiteness represents a particular and hegemonic gaze which although covered, plays a crucial role by normalizing the subordinated position of other ethnicities, often pushed the peripheral position. Leisure, in such a context, far from being an emancipator vehicle towards freedom, re-affirms the whites' hegemony.

"Whiteness becomes an all-pervasive instrumentality, which, like capitalism, threatens to consume the entire world. The existence and survival in the Academy of counter-narratives of race, predicated on communicative rationality, shows that the Enlightenment-while flawed in history-remains a durable ideal of free inquiry" *(Spracklen, 2013, p. 7).*

Following this line of reasoning, one might think that whiteness not only conserves its own rationality but also is replicated through the cultural background of leisure consumption. For this, it is safe to say that over centuries, imperial powers have alluded to ethnocentrism to legitimate possession of strategic lands they coveted. At a closer look, the archetype of leisure occupied a central position marking the borders between civilized citizens and barbarians. To some extent, imperialism, and leisure were inextricably intertwined. Of course, science, and the derived scientific explanations gave universal nature to the prejudices and stereotypes unfolded through the would-be inferior "Others." Not only during the days of social Darwinism which illuminated the glory of British Empire, but also in the hegemony of Roman Empire, leisure, and science pivoted to control the *undesired Other.*

As this argument is given, Spracklen argues convincingly that racism has never came intro fruition without the previously created racial hierarchies which systematically classified the desired racial elements from the undesired ones (outsiders). Such a process of marginalization was culturally normalized through different dispositifs and instruments. The circulation of narratives and stereotypes evinces different power relations as Spracklen remarks:

> *"The 'social effects' of such practices can be very damaging. The power of the stereotype extends to the group that is being stereotyped; because of the asymmetrical power relations, their ability to define themselves is constrained by the dominant discourse"*
> *(Spracklen, 2011, p. 16).*

In this way, the dominant discourse constructs a privilege center, a type of idealized uphill city that never seems to be neutral. Symbolically speaking, a sense of we-who-live-here is fitted against they-who-live-there. Leisure practices are simply hegemonic because they emanate from the will of central authority to remind the citizens who important is to belong to civilized community. Last but not least, the political struggle for representation in leisure spaces is usually accompanied with history tergiversation allowing the imposition of an ideological message reminding the exemplarity of whites over other ethnicities. For Spracklen, heritage tourism offers a fertile ground as example that explains how leisure can be politically manipulated in order for the interests of status quo to be protected. This raises the question to what extent tourism and leisure consumption can be compared alike or studied as similar phenomena.

> *"What can be seen in the literature on tourism is an awareness of the unequal encounter between local sand travelers and the spaces in which those encounters take place- and the purpose of the more commodified end of the industry as sites of residual, hegemonic white, Western culture... the success of heritage tourism of the American West-the cowboys and the frontier towns, recreated in the present day-is a symbol of the fear of white Americans about their future status in the United States. White Americans visit the American West to see a world that is populated by white people fighting and defeating Native Americans ('Indians'), a world where white people are superior and successful in carving out spaces that are free from troubles for themselves. White American tourists can be sure of their whiteness and their ownership of the country through constructing myths of cowboys and frontier life" (p. 35).*

In a nutshell, the main point of critical entry in this discussion suggests that even if *whiteness* may be very well confronted and contested in the popular culture, leisure hot-spots and the industry of tourism, no less true seems to be that popular culture serves as a commodity so that the power of whites was not placed in dispute. This consists in racializing class relations while the real interests of white elite are hidden or transformed as a major dilemma to be embraced widely by the society all. In fact, the success of ideology depends on the capacity to have the rest of society believe the proper cultural values, urgencies, needs, and interests are passively shared, accepted, and internalized as priorities by other classes or ethnicities.

1.5   COMMONALITIES BETWEEN TOURISM AND LEISURE

In the previous sections, we have discussed in depth the notion of leisure, but this begs a more than an interesting question: why are we addressing leisure when the book is focusing on tourism?

In years, an interesting and ever-growing body of studies focused on the intersection of tourism and leisure. Most certainly, scholars agree that tourism is based on the needs of tourists to behave differently than in their original societies. Given this, tourism, which allowed an escapement from the ordinary life to a newly imagined world, should be framed under the sphere of leisure (MacCannell, 1976; Norris and Wall, 1994; Thornton, 1995; Shaw and Williams, 1994; Ryan et al., 1996). As Krippendorf eloquently observes, tourism should be understood as a form of leisure culturally enrooted in mainstream cultural values of the community. Tourism is not good, not bad, only an instrument conditioned by the foundational values of human groups (2010). To a closer look, other voices show an alternative viewpoint arguing that the hedonistic behavior precisely differentiates tourism from leisure (Leontido, 1990). Somehow, as Neil Carr (2002) puts it, the terms tourism and leisure were inevitably entwined. There is a tourism-leisure continuum that needs to be investigated. Just after MacCannell's book which was widely influential in tourism literature, tourism, and leisure were conceived as indistinguishable entities. Although with the same element in common, *hedonism,* tourism, *and* leisure have evolved towards different directions. These differences depend on the behavior of tourists, which is pleasure-oriented to *an animated non-ordinary culture*, where everything can be possible.

To a closer look, tourists often move long distance to pretend to be, what in reality they want to be. This reversal in the role has been exhibited as an observable hallmark of tourist places. The same point has been widely illustrated in the introductory chapter, in authors as Krippendorf or MacCannell who defined tourism as a sub-field of leisure studies. Here two assumptions should be done. On one hand, watching television (a leisure activity) seems not to be the same than other activities at the beach. Here the figure of movement to behave differently than origin is vital to understand the gap between tourism and leisure. On another, no less true is that not all tourists are motivated to make activities which are impossible in their homes. Through a residual culture, which mediates between tourist culture and the routine, some tourists keep the same behavior even on holidays. Equally important, Carr acknowledges that:

> *"The current literature also provides confirmation of the existence of a tourist culture that encourages people to behave in a hedonistic manner to a degree that may not be acceptable in their place of origin. However, there is evidence that not all tourist behaves hedonistically to the same degree and that some actually exhibit behavior that is more restrained and less oriented towards over pleasure seeking than in their place of origin. These findings suggest not everyone is influenced by the tourist culture to the same extent."* (Carr, 2002, p. 980).

In a nutshell, Carr's paper opens the doors to a more than interesting landscape, which unfortunately was not continued. His assumptions revolving around the residual culture indicate not only that people take never a full distance from their original values and rules, even when they are on movement or abroad. Secondly, there seems to be a strong interaction between residual and tourist culture that requires further empirical research. Still further, citizens are falling prey-as passive victims-of the marketers who often manipulate their desires to offer a unique and standardized product. Though consumers are kindly invited to buy an outstanding experience, at the bottom, these experiences are externally designed in a homogenous way. The end of the 20th-century witnesses the rise of radical texts which confronts with the modern leisure as a panacea from where the modern man emancipates himself (Debord, 1998; Virilio, 1999; Roberts, 1999; Kellner, 1999). Not only the process of globalization makes this possible but also the cultural industries introduce the needs of feeling unique experiences and sensations as a new form of relationship (where Otherness is oppressed and neglected). This moot point will be finely discussed in the next section.

1.6   LEISURE PRACTICE, CONSUMPTION, AND GLOBALIZATION

Leisure is a double-edged sword. It is often manipulated by politicians in order to keep the masses under control, but at the same time, leisure works as an instrument of resiliency and enhancement. This ambiguous nature, as Troy Glover (2015) claims, transforms the sense of place paving the pathways for the rise of social change. From Ancient Rome to date, it is not surprisingly that governors devoted their resources to celebrate *games, spectacle, and mega events* in order for gaining further legitimacy in society. These spaces were liminal in nature and everything could happen. Some revolts, conflicts or riots were disparaged during the Gladiator games. There are a lot of insurgent possibilities during leisure time, which needs to be investigated. Glover acknowledges that leisure allows innumerable ways of negotiating the public life, while the ruling class packages and disseminates an ideological narrative to the community.

> *"Leisure is complicit in abstraction. Its seductive guise as unassuming fun and games masquerades its very real function in fashioning tastes and habitual appropriation as cultural capital. Far from trivial, leisure's growing conflation with consumption practices only exacerbates its effects" —(Glover, 2015, p. 102).*

Over the recent years, Glover alerts, governments-associated to the private capital-encouraged the construction and expansion of leisure industries in the city. Through the articulation of *spectacular consumptive environments* Western nation-states conceived leisure as an idealized interpretation of urban life, and doing that they redefined the consumers' minds to a type of recreational consumerism (Glover, 2015). In their introductory section of the book *Landscapes of leisure*, Sean Gammon and Sam Elkington (2015) question the imposition of a hegemonic industry of entertainment built to mediate between people and their environment. Today's leisure is mediated by a bunch of external (fabricated) experiences which draw the contemporary life and of course the sense of place-ness.

"To some theorists, specific spaces and places become increasingly irrelevant. Here the argument is that personal relationships (to places as well as to other people) become less stable, and that more that personal experience and social relations become mediated by information and communication technologies, and thus are disembedded from their local context" (Gammon and Elkington, 2015, p. 2).

The meaning of leisure, as authors adhere, is probably the most difficult object to grasp. In its history, mankind has evolved through the capacity to occupy a position in a specific territory. The sense of space was present in the social sciences as the conceptual touchstone of territoriality, but not these stable forms are setting the pace to more hybrid meanings. The place is constantly disputed, negotiated, and affirmed according to each agent's biography. They coin the term *leisure-scape (leisure landscape)* to denote the articulation of flowing bodies, perception, places, and identities mainly associated with modern leisure practices. One might speculate that the notion of landscape evokes the idea of *something out there*, but indeed the idea of landscape is lived and internally reminded in the mind. David (2015) employs the metaphor of flirting to define leisure. In his viewpoint, like flirting spaces of leisure are embodied in many ways, engaging with the uncertainness of being alive in the world. Leisure spaces-far from being established entities-are not controlled by the exercise of power or externally fixed. Contrariwise, they move highly contingent. Leisure has two important components which constitute it: the gaze and design. While the former signals to the setting or features that allows leisure practices, the latter reserves to *the landscape formation.*

Thus, in cultures of doing leisure, cultures can be emergent through leisure, along with particular ways of thinking landscape, and place, feed or are fed, into individuals' and group's identity, through their own participation rather than as directed by particular reading of significations due to outside frameworks" (Crouch, 2015, p. 14).

Doubtless, Crouch triggers a new and hot debate revolving around the pre-figuration of leisure research today. We are accustomed to studying leisure practices through objects or building, temples, sports stadiums, shopping malls, parks, games, or other activities which can be gazed. Instead, leisure landscapes mark the beginning of a more fluid dynamic where the self-connects with others forming and shaping its own identity. This process is never fixed, but contingent and liminal, if not subject to many risks and challenges (even so threatening us). In consonant with this, Sam Elkington (2015) elaborates a more than an interesting diagnosis. The post-modern urbanism is gradually creating new forms of leisure practices, which far from being alienating, changes social relationships. He cites the example of a young boy connected in his smart-phone in a park, inter-rogating to what extent he is living a real life? Of course, Marxist theory punctuates on modern leisure as subject to the force of the free market, commoditizing the social relation to the power of capital. However, there are many studies which focused on the resistances and crevasses in leisure spaces of everyday life. As he puts it:

> "This contemporary narrative of leisure place, as we shall see, has legitimate provenance as a way of looking, an orientation, to social-spatial manipulation specifically because of the way it accommo-dates the multiplicity of personal narratives of individual leisure experiences. In the context of mass leisure, an alternative narrative of complexity, connectivity, and transformation is put forward as a way of encountering and exploring everyday urban space on its own terms and of opening up to the reality that, contrary to what many detractors believe, today's urbanized society is still creating real human places" (Elkington, 2015, p. 24).

Having said this, modern leisure practices not only illuminate in new directions the classic readings overlooked but also enriches the human existence in ways scholars do not understand. Rethinking the leisure places invites us to imagine an emerging conceptual corpus where the tension between places and non-places simply dilutes. Placing Marc Augè's theory under the critical lens of scrutiny, Elkington suggests that the idea of *non-places* as commoditized spaces of consumption and deper-sonalization where the individualism prevails, only tells us a part of the story. The European *intelligentsia* was historically prone to understand the daily habits as constrained to a place, which is a symbolized space, left to a specific (forged and negotiated) history and collective identity. However, other voices have criticized this position emphasizing furtherly on the movement-instead of the place. Starting from this premise, Elkington argues convincingly that though the term *non-place* seems to be trouble-some, what is more, important is to pay attention to the leisure place as a place where people do something, instead of a place where something is. Such a conception is crucial to overcome the methodological problems leisure studies have in this hyper-globalized world.

1.7   CONSIDERATIONS FOR A FUTURE AGENDA IN LEISURE STUDIES

As debated in previous sections, not only leisure studies are facing serious challenges because of the rise of new digital technologies which are altering the nature of leisure but also appears to be in crisis (Rowe, 2002; Bramham, 2006; Henderson, 2011; Blackshaw, 2014). As Coalter (1997) eloquently highlighted, although the growth of scientific publications, journals, books, Ph. Doctorate dissertation theses and the recent interest for other consolidated disciplines to understand leisure, scholars have failed to explain the cultural and individual meanings linked to the phenomenon. Based on a positivist approach, these studies overlooked the connection between leisure and culture probably misjudging leisure meaning. Therefore, he bemoans the introduction of more qualitative insight to confront with the current paradigm. In this vein, Cara Aitchison (2010) suggests that the origin of this epistemological crisis stems from the reluctance of researchers to articulate the recent advances from other disciplines into a clear model that explains the issue. Of course, as she understands, this happens simply because of some scholars rejected the post-structuralism as a fertile ground to forge a critical discourse against the established dominant narratives. As Aitchison notes, post-colonial feminism, as well as post-structuralism, should be embraced as an all-pervading conceptual model that helps to understand the radical shifts leisure studies have recently experienced. Basically, a poststructuralist's advance, which plays a crucial role in deciphering the dark side of power and politics, has been marginalized towards a peripheral position within Academia. In a seminal paper, which captivated our attention, David Rowe (2002) argues that the crisis of leisure studies has multiple reasons and consequences. Almost 20 years after this publication his findings are still valid. Per his viewpoint, the current crisis depends on the rivalries among academic groups who struggle to impose their own meaning of the term. It is important not to lose the sight of the fact that if leisure pattern changes, the paradigms changes. This seems to be what the crisis in the fields leisure studies represent, a lack of understanding of the new forms of leisure consumption in the 21ˢᵗ century. He goes on to write:

> *"Current defenses of the field of Leisure Studies tend to be institutionally rather than intellectually founded. That is, for perfectly understandable reasons they want to protect Leisure Studies, not as delimited social scientific inquiry so much as an intellectual institution (or, in Foucault's terms (1972), as discursive formation). It does not or cannot, therefore, recognize the possibility that Leisure Studies may not be what it used to be because either the conditions under which it is practiced have changed or, more radically, that the object of analysis has, indeed, changed." (Rowe, 2002, p. 7).*

In recognition of the above-cited excerpt, Karla Henderson (2010) acknowledges that the 21ˢᵗ century was the turning point to the crisis of leisure studies. Most certainly, scholars manifest often their worries for the fragmentation of the discipline, even for the countless definitions of leisure circulating everywhere, but this is not the main reason behind the current crisis, as Henderson adheres. Over the years, leisure studies remained linked to their original position emerged in the US where leisure practices were mainly understood as practices of recreation and entertainment. This position was gradually shifting towards more politicized definitions. Those American scholars, who struggled against the economic-based paradigm who only trained students to work in recreational parks, emphasized on the idea that considering leisure as a social institution, to be more exact to be a question of social identity. In consequence, the rise and subsequent expansion of postmodernism allowed the coexistence of numerous identities which invariably ushered the discipline in a state of unquestionable stagnation (if not fragmentation). Doubtless, the main challenge for the next years is to fix clearly what is the mission and value of leisure studies. Last but not least, in recent books we have stressed the urgency and the needs of re-conceptualizing not only leisure as main object of study but also tourism in view of a darker form of consumption-oriented to gaze the Other's pain. This new type of consumption widely practiced in new segments as dark-tourism, war-tourism or disaster-tourism is centered in a new version of capitalism where the Other's death is the main commodity to exchange. In a world that resists disappearing, the Other's death is the only reason for self-enhancement and happiness (Korstanje, 2016, 2018). The present chapter, as well as this book, aims at discussing the nature of leisure and tourism while focusing mainly on the radical changes the discipline went through over the recent decades. Such a point will be dully addressed in the next chapters.

1.8   CONCLUSION

As originally stated, we have looked to bolster a conceptual dialog between leisure studies and other social sciences as sociology and anthropology. Of course, this dialog or inner-exchange should not be conditioned to the advances of sociology and anthropology alone. At a first glimpse, these disciplines pivoted the earlier attempts for deciphering the nature of leisure and for that, at least, deserve some recognition. We cannot understand leisure without touching Erich Weber or Norbert Elias. The first section discussed initially the role of leisure in society from the lens of Frederick Munnè, a Spaniard cultural theorist who has devoted his life in the study of leisure. Echoing Weber and other theorists, he constructed a bridge between cultural and leisure studies. Secondly, Munnè's legacy facilitates us to adventure towards more sophisticated texts and authors as Veblen, Weber, or Elias. Although each author maintained his position, a common argumentation can be found: leisure acts as a social institution oriented to keeps society united while the institution working. To put the same simply, leisure occupies a central position in the construction of the social scaffolding. In third, we question to what extent leisure and tourism can be compared or treated in isolation. Scholars, over years, have debated the commonalities and differences between tourism and leisure. As a sub-form of leisure, tourism revitalizes the socio-psychological frustrations of citizens during their working time. Citing MacCannell or Urry, tourism paves the ways for the correct functioning of social institutions. The last two sections are reserved to the expansion of globalization, as well as the rise of cultural industries that venerate leisure as a good practice and the current crisis leisure studies experience today. The main thesis held here punctuates that leisure practices are being shifted to a darker version where the Other's suffering is the main criterion of attraction. This point, which was illustrated by this author in his earlier books, is known as *Thana-Capitalism.* Needless to say, that the concept is carefully described in the next chapters that form this editorial project.

KEYWORDS

- anthropology
- bourgeois leisure
- etymology
- history of consumption
- hyper-globalized
- leisure
- research
- tourism

REFERENCES

Aitchison, C., (2000). Post structural feminist theories of representing others: A response to the 'crisis' in leisure studies' discourse. *Leisure Studies, 19*(3), 127–144.

Balme, M., (1984). Attitudes to work and leisure in ancient Greece. *Greece & Rome, 31*(2), 140–152.

Blackshaw, T., (2014). The crisis in sociological leisure studies and what to do about it. *Annals of Leisure Research, 17*(2), 127–144.

Bramham, P., (2006). Hard and disappearing work: Making sense of the leisure project. *Leisure Studies, 25*(4), 379–390.

Carr, N., (2002). The tourism-leisure behavioral continuum. *Annals of Tourism Research, 29*(4), 972–986.

Coalter, F., (1997). Leisure sciences and leisure studies: Different concept, same crisis? *Leisure Sciences, 19*(4), 255–268.

Crouch, D., (2015). Unravelling space and landscape in leisure's identities. In: Gammon, S., & Elkington, S., (eds.), *Landscapes of Leisure* (pp. 8–23). Palgrave Macmillan, London.

Debord, G., (1998). *Comments on the Society of the Spectacle* (Vol. 18). Verso, London.

Dumazedier, J., (1967). *Toward a Society of Leisure*. Free Press, New York, NY.

Dumazedier, J., (1999). The hidden importance of increased free-time in civilizational change. *Loisir & Societe-Society and Leisure, 22*(2), 313–322.

Elias, N., & Dunning, E., (2008). *Quest for Excitement: Sport and Leisure in the Civilizing Process*. University College Dublin Press, Dublin.

Elkington, S., & Gammon, S., (2015). Reading landscapes: Articulating a non-essentialist representation of space, place and identity in leisure. In: Gammon, S., & Elkington, S., (eds.), *Landscapes of Leisure* (pp. 1–7). Palgrave Macmillan, London.

Elkington, S., (2015). Disturbance and complexity in urban places: The everyday aesthetics of leisure. In: Gammon, S., & Elkington, S., (eds.), *Landscapes of Leisure* (pp. 24–40). Palgrave Macmillan, London.

Glover, T., (2015). Animating public space. In: Gammon, S., & Elkington, S., (eds.), *Landscapes of Leisure* (pp. 96–109). Palgrave Macmillan, London.

Henderson, K. A., (2011). Post-positivism and the pragmatics of leisure research. *Leisure Sciences, 33*(4), 341–346.

Kellner, D., (1999). Virilio, war and technology: Some critical reflections. *Theory, Culture & Society, 16*(5, 6), 103–125.

Kelly, J. R., (2012). *Leisure* (4ᵗʰ edn.). Sagamore Publishing, New York, NY.

Kelly, J. R., (2019). *Freedom to be: A New Sociology of Leisure*. Routledge, Abingdon.

Korstanje, M. E., (2009). Reconsidering the roots of event management: Leisure in ancient Rome. *Event Management, 13*(3), 197–203.

Korstanje, M. E., (2016). *The Rise of Thana-Capitalism and Tourism*. Routledge, Abingdon.

Korstanje, M., (2018). *Tracing Spikes in Fear and Narcissism in Western Democracies Since 9/11*. Peter Lang, Oxford.

Krippendorf, J., (2010). *Holiday Makers*. Routledge, Abingdon.

Kyle, D. G., (2012). *Spectacles of Death in Ancient Rome*. Routledge, Abingdon.

Leontido, L., (1994). Gender dimensions of tourism in Greece: Employment, sub-structuring and restructuring. In: Kinnaird, V., & Hall, D., (eds.), *Tourism: A Gender Analysis* (pp. 243–249). Griffith University Press, Australia.

MacCannell, D., (2013). *The tourist: A New Theory of the Leisure Class*. University of California Press, Berkeley, CA.

Munné, F., & Codina, N., (2002). Ocio y tiempo libre: Consideraciones desde una perspectiva psicosocial. (Leisure and free time: Some questions from a social psychology perspective). *Revista Licere-Brasil, 5*(1), 59–72.

Munné, F., (1980). *Psicosociología del Tiempo Libre: Un Enfoque Crítico (Psychosociology of Leisure: A Critical Approach)*. Trillas, Mexico.

Norris, J., & Wall, G., (1994). Gender and tourism. *Progress in Tourism, Recreation and Hospitality Management, 6*, 57–78.

Pratt, M. L., (2007). *Imperial Eyes: Travel Writing and Transculturation*. Routledge, New York, NY.

Roberts, K., (2006). *Leisure in Contemporary Society*. CABI, Wallingford.

Rojek, C., (2005a). An outline of the action approach to leisure studies. *Leisure Studies, 24*(1), 13–25.

Rojek, C., (2005b). Leisure theory. *Principles and Practice*. Springer, New York, NY.

Rojek, C., (2013). *Capitalism and Leisure Theory (Routledge Revivals)*. Routledge, Abingdon.

Rowe, D., (2002). Producing the crisis: The state of leisure studies. *Annals of Leisure Research, 5*(1), 1–13.

Ryan, C., Robertson, E., Page, S. J., & Kearsley, G., (1996). New Zealand students: Risk behaviors while on holiday. *Tourism Management, 17*(1), 64–69.

Shaw, G., & Williams, A. M., (1994). *Critical Issues in Tourism: A Geographical Perspective*. Blackwell Publishers, New York, NY.

Spracklen, K., (2011). *Constructing Leisure: Historical and Philosophical Debate*. Palgrave Macmillan, Basingstoke.

Spracklen, K., (2013). *Whiteness and Leisure*. Springer, New York, NY.

Stebbins, R. A., (2011). Leisure studies: The road ahead. *World Leisure Journal, 53*(1), 3–10.

Thornton, P. R., (1995). *Tourist Behavior on Holiday: A Time-Space Approach*. Doctoral dissertation, University of Exeter, UK.

Toner, J. P., (2013). *Leisure and Ancient Rome*. John Wiley & Sons, New York, NY.

Veblen, T., (2017). *The Theory of the Leisure Class*. Routledge, Abingdon.

Virilio, P., (2002). *Ground Zero*. Verso, London.

Weber, E., (1969). *El Problema del Tiempo Libre. (The Problems of Leisure)*. Editorial Nacional, Madrid.

Wynne, D., (2002). *Leisure, Lifestyle and the New Middle Class: A Case Study*. Routledge, London.

CHAPTER 2

Towards a Theory of Tourism

ABSTRACT

Over years, sociology and anthropology have systematically overlooked the important role played by leisure to regulate the different conflicts and cleavages happened in society during the working days. A young Norbert Elias acknowledged the importance of leisure not only revitalizing social frustration but socializing the mainstream cultural values of society. This chapter interrogates furtherly on the different academic waves which have focused on the problem of modern leisure. From different angles, each one has limitations and notable advances in the field. We start the chapter with a historical and conceptual insight on ancient leisure. The efforts devoted to this chapter aims to validate that tourism should be understood as a form of leisure. In the successive sections, we dissect critically the main theories in leisure research laying the foundations to a new agenda for the years to come.

2.1 INTRODUCTION

In 1968 the International Union of Official Travel Organizations (known now as UNWTO, World Tourism Organization), which was the major entity in the United Nations specialized in tourism policies and development strategies, defined tourism as *the activity that comprises activities of persons traveling to and staying outside their usual environment for not more than once a consecutive year for leisure, business, and other purposes.* Although this above-cited definition sheds some light generation after generation of countless tourism-related researchers, as well as policymakers, no less true is that over the recent decades, scholars not only disputed the authority of UNWTO to define tourism but also questioned radically such a definition as originally ingrained in a quest for profit (Craik, 2002; Jennings, 2001; Theobald, 2005). In fact, UNWTO has triggered an economic-centered paradigm seeing tourism as a growing industry that recycles, develops, and boosts local economies. Like in leisure studies (a theme discussed in the first chapter), one of the limitations of tourism research alludes to a much deeper crisis that accelerated the discipline stagnation. As John Tribe puts it, not only the economic-based theory as well as management has mined the Academia inside, but also there is a clear dispersion of academic waves which gradually pushed the applied research to an indescribable state of chaos which he dubbed as *"the indiscipline of tourism research."* Paradoxically, after decades of fine and robust research and debate, the multiplication of books, journals, and Ph. Doctorate dissertation, less is known about the origin and evolution of tourism. For Tribe, the problem becomes more acute since the Academia historically failed to create an all-encompassing theory to understand tourism that the rest of scholars follow (Tribe, 1997, 2000, 2010).

As this backdrop, the meaning of tourism has created great controversy within the academic circles. At least three different epistemological traditions have not agreed with the etymological nature of the term tourism. Neil Leiper traces back the origin of tourism to the word *Grand Tour* which derives from the name of the first family *(de la Tour)* which managed business travel between England and France. This event took place in the middle of the 18th century and was considered a foundational point for the rise of shopping travelers (Leiper, 1979, 1983). Contrariwise, the Anglo-Saxon school suggests that the term comes from the Old Saxon term *Torn* which means going around. From this term derives *Torn-are (going around) and Torn-us (what is going around)* (Fernandez Fuster, 1967). As Fernandez Fuster suggests, the word was often used by the medieval peasants in England to denote those travels marked by the intention of return. But things come worse to worst; Arthur Houlot (1961), who is the founding parent of Semitic school, proposed that the origin of tourism should be found in the Old Aramaic term *Tur* used for the explorations of Moses to Canaan in the Old Testament. In some perspective, the dispute-far from being closed-characterized the knowledge fragmentation prevailing today in tourism research (Fernandez Guzman, 1986; Korstanje, 2007). Of course, the question of whether the etymology of tourism remains unclear, the same applies to the definition of tourism. Hence, it is essential to note that academicians divide into two clear-cut waves. On the one hand, some voices claim that tourism is a modern industry emerged after the Industrial revolution by the avocation of complex processes which oscillated from the improvement of working condition and the technological breakthrough in the constellations of transport and mobility. For these authors, it is safe to admit, tourism, and economy are inextricably intertwined. For them, tourism should be framed as a productive activity. Of course, as they agree, even if tourism accelerates undesirable consequences for the local economies such as contamination, gentrification, or real-estate speculation, through the articulation of rational planning, policymakers may find and eradicate those risks that jeopardize the sustainability of the industry (Medlik, 1991; Archer, 1991; Frechtling, 1991; Van Doorn, 1991; Likorish and Jenkings, 2007; Lee and Chang, 2008). What is more important, this academic wave, which accompanied the discipline from its inception (since 70s decade to be precise), not only triumphed in expanding a managerial vision to universities and journals but also homogenized the formation of Curricula in tourism-related careers and courses (Amoah and Baum, 1997; Airey and Tribe, 2006; Okumus and Yagci, 2006). The academic-based paradigm replicated ideologically through the tourism educative system. At a closer look, today's universities train future tour-guides and professionals to be rapidly inserted in the industry instead of scientific researchers or social theorists (Dann and Cohen, 1991). On the other hand, a reactionary wave, which surfaced in the 70s, headed by sociology and anthropology, emphasized the needs of studying tourism as a social institution designed to regulate the internal social frustration while revitalizing *the social bondage.* For these theorists, tourism should be considered as something else than a mere industry. The advance of industrialism and capitalism have been eroded the traditional relations as well as the classic performance of institutions creating radical shifts in society. In this vein, tourism would be approached as a catalyst instrument that revitalizes and regenerates the psycho-sociological frustration of the modern worker. In a nutshell, tourism should be framed as a subtype of leisure practice (MacCannell, 1973, 1976, 2002; Cohen, 1988; Dann, 2002; Salazar and Graburn, 2014; Urry, 1990, 2002; Meethan, Anderson, and Miles, 2006). While the economic-based theory signaled to management and marketing as guiding discipline to design the tourist destination, the social-led paradigm devoted considerable efforts to show and explain the conflict relations between hosts and guests as well as the social maladies locals often face which are recreated by the tourism industry (Aramberri, 2001; Tucker, 2003; McNaughton, 2006). In her landmark book, *Host,* and *Guest: the anthropology of tourism,* Valene Smith (1989) argues that anthropology has the mandate to guide the next steps in tourism research, collaborating with other disciplines, but calibrating the focus on the social nature of tourism. Per her viewpoint, it was true that the original definition unilaterally imposed by the UNWTO was markedly impregnated of some methodological limitations and centered on an ethnocentric discourse, a problem scholars should correct. Quite aside from this controversy, the present chapter provides a snapshot of the theories and main authors who have attempted to define what tourism is.

Let us remind readers how the structure of this chapter was finally organized. The first section is limited to the advances and contributions of different scholars in the field of tourism management. Based on the findings of the main management theorists, the section explains not only what are the factors that were a full-fledged part in the configuration of the tourist industry but also helps conceptually to understand the modern tourism industry. Secondly, we review the bibliography of those social scientists who have theorized on tourism and its effects for the community. By this way, the motivations of tourists to select some destinations while avoiding other ones, as well as the psychological profiles of tourists, are carefully assessed in the final section.

2.2   TOURISM MANAGEMENT AND TOURIST DESTINATION

Doubtless, as discussed in the earlier section, Management, and Marketing have pivoted the advance of tourism research over the two last decades (Buhalis and Costa, 2006; Page, 2007; Mountinho and Vargas-Sanchez, 2018). In this part, we shall dissect the main contributions and limitations of managerial disciplines applied to tourism fields. Here two significant aspects should be discussed. On the one hand, authorities, and policy-makers valorize the importance of tourism as a vehicle towards prosperity and economic growth (Pearce, 1982). On another, as debated in the second chapter, scholars' toy with the belief that tourism and leisure practices are characteristic of democratic and laissez-faire societies (Munnè and Codina, 2002). In this token, tourism is seen-from its inception-as a useful mechanism of inter-cultural understanding and respect that gradually promotes prosperity and peace worldwide (Raymond and Hall, 2008; Richards, 2018). From this moment onwards, paradoxically, the discipline reaches an irreversible point of stagnation where the quantification of travels and tourism consumption marked the political turf of tourism-related researchers. Stating it unpretentiously, more tourism entails more democracy and prosperity and vice-versa. Tourism management launched to colonize and monopolize the tourism knowledge production, but in so doing, the discipline divided in two clear-cut waves: scholars who are interested in understanding and predicting the psycho-social motivation of tourists and those who focused on demographic and structural features of tourist destinations to offer a catch-all concept of the term (Isoahola, 1982; Mannell and Isoahola, 1987; Buttler, 1991; McCabe, 2001; Pearce and Lee, 2005). One of the pioneers who have devoted his time in this direction was Douglas Pearce. Per his stance, the psychology of tourism offered a fertile ground to grasp the complex motivations of tourists as well as its intersection with the dichotomies of the tourism industry. The tourist-as an agent-never moves alone, he or she is culturally wedded to an environment, the industry and his or her internal emotions. For Pearce, the main goal of tourism research should propose a coherent model to reconstruct how tourist behavior evolves in different conditions (Pearce and Butler, 1993; Pearce, 1991, 1999, 2012). In the preface of his book *Managing tourism,* Medlik (1991) overtly says that beyond the psychological motivations which determined tourism behavior, the tourism industry showed *a remarkable resiliency* not only to face external threats as terrorism, political violence, which may place the industry in jeopardy (Richter and Waugh, 1991) but also in regulating the negative effects of the activity in the environment. Therefore, tourism management illustrates policymakers to promote sustainable campaigns and programs oriented to revitalize-if not protect-the tourist destination. A new academic trend is now emerging, a new one concerned by the well-functioning of a tourist destination (Medlik, 1991; Fretchling, 1991; Jenkins, 1991). Stepping back to the motivation and the tourist destination theory, one might speculate that both theories were oriented to understand the industry but originally underpinned in two different needs or questions: *why do tourists travel? And why do some destinations rise while others simply decline?*

Here some reflections on what is the correct method to understand the international demand should be more appropriate. For example, in 1985, Uysal and Crompton publishes a more than interesting manuscript at the Journal of Travel Research, one of the leading journals of tourism fields. In this text, they hold that a correct decision to invest in a destination depends on the capacities of policymakers to reach the right diagnosis, which is crucial for predicting the fluxes of the tourist demand. Authors conclude that statistical instruments have advantages but at the same time, some disadvantages. Hence, qualitative methods should be alternated to quantitative ones in order to get a correct diagnosis in the evolution of the demand in an ever-globalized world. Scholars and policymakers were working hard to identify the psychological variables that condition the decision-making process as well as the needs of traveling abroad. The theory of roles, needless to say, gave a clear solution to this dilemma.

One of the authoritative voices in the fields of psychology of tourism applied to management issues is Stanley Plog. Subsidized by some important airline companies, he develops a model based on three main variables. The Plog's paper *Why destination areas rise and fall in popularity* sees the light of publicity in 1973. In this seminal paper, he suggests that our cognitive condition to perceive the surrounding environment is invariably interlinked to our emotionality. In view of this, some reactions to feel attracted or threaten by some situation depends upon the agent's emotional inner world.

From this perspective, any tourist experience can be explained by the biography of the subject as well as the role and its respective personality (character asset). Following this, he classifies tourists according to a continuum of three variables: (a) people who previously educated in contexts of sociability and kindness have developed a prone to the Other (allocentric personality); (b) those who socialized in a home of conflict have some problems to be open to the Other and looks for safe places (psychocentric personality); and (c) and mix of the other two assets where the involving person alternates states of closure and openness to new situations and cultures (mid-centric personality). Though criticized by the lack of scientific rigor of his study (Hoster and Lester, 1988), Plog lighted the torch for what other scholars have taken up when the theory of risk perception is adopted as a mainstream paradigm just after 2001. A decade later, in 1985 to be more precise, Douglas Pearce (1985) conducts research describing a sample formed by 100 interviews, in which case, he found 15 different roles. Each segment is grouped according to different reactions and behavior while traveling. Over decades and from different angles, Management research systematically emphasized on the dichotomies between pleasure and business towards the creation and full evaluation of different typologies associated with tourists' roles. The main common-thread argument of these studies agrees that those tourists who travel for business are less interested in knowing new places and people, while those who travel for pleasure purposes are more open to new cultures and customs (Cohen, 1972, 1974; Snepenger, 1987; Yiannakis and Gibson, 1992; Prayag, Hosani, and Odeh, 2013). Once Plog responded why people are motivated by travels, researchers devoted their resources to forecast and regulate international demand. A key explanation to better understand this is originally given by the role of destination-image which is defined as a mental process centered in the fluid interaction of cognitive image, the gathered information and the experience which is biographically constructed (Reynolds, 1965). One of the most influential researchers in the fields of the destination image is John Crompton. In different stages of his career and works, he acknowledged that the subject-at the time of deciding where to travel-is involved in five distinguishable stages: (a) the rise of stereotypes and discourses that lead to imagine the destination; (b) the benefits and limitations of choosing a specific tourist destination; (c) the interplay between those destinations in mind (imagined) and those evoked (which means those where the agent was already visited; (d) the active quest of information about the marked destinations; and (e) the final decision. Jointly to Um, Crompton elaborates the model of *push and pull factors,* a work that tasted the proof of time and was widely cited in the management tourism research. The idea of push factors is shaped by the internal drives that move tourists to seek new activities escaping from their routine. The pull factors can be explained as the necessary forces created by the destination to attract these tourists. While push factors are the needs of relaxing or the escapement, or the quest of prestige and distinction, pull factors appear when the destination features captivate the attention of the demand (Um and Crompton, 1990). Undoubtedly, the push and factor theory not only were replicated in countless studies across the globe (Fodness, 1994; Uysal and Jurowski, 1994; Gnoth, 1997; Cheng and Hsu, 2000; Lee et al., 2002; Jeong, 2014; Rice and Khanin, 2019), but also situated as the tug of war of tourism management research even to date. It is important to say that this section does not reflect all published works and studies in the constellations of tourism management but it is only limited to the first steps and evolution of research towards a consolidated sub-discipline of tourism. Some of the critiques directed by the sociology of tourism to management associate to the lack of interest in understanding the issue-perhaps as a social institution and its relation to society-unless by the profits or business tourism daily generates. More interesting in protecting the industry than in studying tourism multidisciplinary, tourism management was torn in a methodological dilemma that gradually led this sub-discipline to a gridlock. In fact, tourism management systematically failed to explain what tourism means or, what is worse, why tourists need to travel. Instead, they planned to trigger the marketing-related campaign or management programs to mitigate the negative effects that threaten the destination image. On behalf of the science or the scientific reason, they implanted an economic-based approach that invariably plunged the discipline into an epistemological crisis as never before (Tribe, 2010; Hall, 2013; Laws and Scott, 2015; Korstanje, 2018). Not surprisingly, the gap left by tourism management research motivated other scholars-who came from social science-to offer an all-pervading diagnosis of tourism.

For Robinson, Luck, and Smith, tourism exhibits a fascinating and dynamic nature which covers a vast range of activities. As a dense net of sub-sectors, tourism sometimes escapes the economic sphere, including travelers, visitors, staff, local hosts and so forth. Nonetheless, there is a clear tension between research questions and tourism management oriented to policy questions. For authors, research questions may very well solve the problems of management but following the scientific method. These management issues are multifaceted and very difficult to resolve. As the authors said:

> *"Social Science research can be directed to examining tourism from a wide range of perspectives such as a form of human behavior with a focus on understanding individuals or as a social phenomenon, with a focus on looking at tourism in a group of social contexts. It usually involves a specific disciplinary focus, such as anthropology, economics, ethnography..." (Robinson, Luck, and Smith, 2013, pp. 454, 455).*

In some occasions, even the solution needs further money, diplomacy, arbitration, and monetary resources. In this vein, social sciences, which are fertile ground for expanding the current knowledge of the industry, is systematically misjudged, or pressed to a peripheral position in tourism-related careers (Robinson, Luck, and Smith, 2013).

2.3   TOURISM AS A SOCIAL PHENOMENON

Most probably because the founding parents of sociology and anthropology developed a pejorative connotation of consumption and tourism was the reason behind the decision of the first sociologists of tourism to continue such a tradition. For Emile Durkheim, for example, the advance of industrialism would inevitably efface the social ties accelerating the rise of social maladies as anomie, alcoholism, and suicide (Durkheim, 1976, 2014). Max Weber understood capitalism as *an iron cage unilaterally* imposed as closed, one-sided to legitimate the rational instrumentality of capitalism while accelerating a process of individuality and depersonalization. Although countless social scientists have been certainly captivated by tourism research, no less true is that the sociology or anthropology of tourism is situated as a bit player in a game monopolized by managerial disciplines.

As the previous argument is given, the sociology of tourism coordinated efforts to describe tourism and its evolution in the threshold of time. One of the pioneering scholars, who has been debated in the introductory chapter, was doubtless Dean MacCannell. Notably influenced by the structuralism of Emile Durkheim and Claude Levi-Strauss, MacCannell holds the thesis that tourism is limited to the rise and expansion of capitalism. Centered on Marxism and Structuralism as main traditions, he toys with the belief that tourism should be historically framed as a modern phenomenon surfaced after the industrial revolution. For this reason, tourism would never be feasible in ancient times. As a combination of technical and economic factors, which paved the ways for industrialism, tourism certainly keeps the society united. Like Durkheim, MacCannell strongly believes that industrialism erodes the tenets of society undermining reciprocity and trust. In the constellations of primitive societies, the figure of totem emanates the authority of chiefdom, as MacCannell adds. Likewise, in modern societies where the totem has no authority, tourism mediates between citizens and their institutions. In other words, the same role played by totem in the primitive cultures is continued by tourism in modern society. It is safe to say that the sociology of MacCannell is marked by four academic waves. From Durkheim, he takes the division between the profane and the sacred space while from Goffman the concepts of back and front stages. Karl Marx gives some insight on class struggle and the role of ideology or alienation insofar Levi Strauss provides him with the conceptual framework of structuralism, above all the dichotomies between the modern and the primitive minds (MacCannell, 1973, 1976). With this background in mind, he proffers a more than an interesting model where tourism fills the gap left by the secularized life. While totem plays a crucial role in primitive organizations not only by mediating between people and their institutions but enhancing social cohesion, tourism revitalizes the psychological frustrations often occurred during the working day. At a first glimpse, consumption, and tourism dispose by the ruling class to control the workforce preventing the social disaggregation but at the same time, creating an interest for novel experiences (MacCannell, 1976, 1984). In opposition to Urry, who holds that the agency elaborates its own relation with structure negotiating a phenomenological position in this world, MacCannell alludes to tourism as a force which recreates dominant and unilateral narratives, discourses as well as stereotypes (finely ingrained in the figure of what he subbed as staged-authenticity) oriented to reproduce *empty meeting grounds* (MacCannell, 2001, 2011, 2012). Here his argument shares a point of convergence with Urry. While touring, cultures, communities, and even people are commoditized according to a cultural matrix that precedes the tourist sight. In consonance with French philosophers as Marc Augé (1995), or Paul Virilio (2006), he eloquently remarks that the late capitalism not only disorganizes human relations but also reproduces the conditions towards *empty spaces* where the ethnic difference is widely homogenized and monopolized by the market. The dominant discourses of West are aimed at marking the "Others," but in so doing, the white elite is unmarked. The sense of freedom given to tourists situates in a privileged position respecting to others who are immobilized. At a closer look, the obsession for consuming "Others" seems to be marked by egoism and individualism. Reluctant to be in contact with other tourists, modern sightseers move as cannibals limiting the real contact with natives beyond the landscapes of the imagined "Other." This represents a real lack he dubs as the eternal lack. The quest for something new is preceded by a much deeper sentiment of guilt, which rests in the aboriginal world's extermination. Since West has never asked for pardon for the Conquest of the Americas, it was enrooted in its consciousness in a pathological mode (MacCannell, 2001, 2011, 2012). Paradoxically, the quest of authenticity leads invariable to depersonalization, and only ethics, as MacCannell adheres, can solve the problem. Although MacCannell sheds light on the sociology of tourism leaving a legacy followed by countless scholars, he was criticized because his observations which apply very well for Disney-world cannot be replicated in other tourism-contexts. This leads John Urry to bolster a critical dialog with MacCannell adopting a Foucaultian term: *the gaze which helps* to describe the Ocularcentrism enmeshed in the Western lifestyle. Through the act of gazing, as Urry writes, the self not only dispossesses "Others" but connects with a cultural matrix that gives sense to what is being seen. At the time this matrix is deciphered, scholars understand how the different gazes and gazers are framed and interlinked. Needless to say, that Urry knows three types of different gazes coexist or even are alternated in the threshold of time: individual, romantic, and solitary.

As he cites, "I call the romantic gaze, solicitude, privacy, and a personal, semi-spiritual relationship with the object of the gaze are emphasized. In such a case, tourists expect to look at the object privately or at least only with significant others" (Urry, 2002, p. 150).

Critically speaking, Urry is not happy with the argument that punctuates tourism is a vehicle of peace and prosperity, but to some extent, his argu-mentation divers from MacCannell. Urry sees the gaze as an opportunity to transform (recycle) spaces fraught of risk and fear, as commoditized and embellished areas of consumption and comfort. But far from being real, these areas are ideologically orchestrated to serve to the interests proper of an *imperial economy.* The power of gaze does not consist in what it dispossesses but in the created gap between gazers and their objects. In an ever-expanding world where institutions are in shaky foundations, the tourist gaze interprets and reformulates a new type of circular logic which he names as circular (esthetic) reflexibility. The current cultural values of postmodern societies confer to their citizens the autonomy to travel elsewhere, any geographical point of the globe. In Urry's insight, tourism, and consumption surged only in modern times. The meaning of tourism depends not only of the features enabled by geographical displacements but (in this Urry coincides with MacCannell) in the intersections of leisure, consumption, and labor. The tourist experience is based on ongoing nego-tiations that changed the ways and how people gaze. Open to the logic of escapement, leisure allows the liberalization of all social constraints, which are fulfilled by the economy of desire (sign) and consumption. Unlike other of his colleagues, Urry contends that tourism is a modern phenomenon associated with consumer services (Urry, 2002, 2007). Having said this so the question is how does this gaze work?

The tourist gaze is based on the trust, which always evinces the needs of controlling the Other. This dispositive of control, originally designed to subordinate the Non-Western "Other," was historically created by the Western civilization since the 18th and 19th centuries. Urry is reluctant-as MacCannell does-to accept that the "Other" is passively possessed by the tourist gaze. He, rather, admits that a new growing context of hyper-mobility has fabricated a new symbolism that connects citizens to their institutions and poor with rich countries. The theory of development not only fell short in reproducing the conditions to boost under-developed economies but also reaffirmed the center-periphery dependency. The rapid evolution of the tourism industry accelerated the growth of the Global North while pressing the Global South to ask for loans. The needs for consuming circulating signs cemented a new logic in the global market where spaces are classified in desired and undesired types.

In his seminal book, *The Tourist Gaze* (Urry, 2002), Urry notes that geographies, persons, and landscapes are carefully organized through gazing. Far from being a disorganized activity, it related to a cultural model that persisted in the time. In order for the economy of signs to expand, mobilities connect psychological needs externally designed with commodities. This is exactly the point of discussion shown in *The Economies and Spaces,* a project which is co-authored with Scott Lash. The current atmosphere of multi-culturalism is unable to find the division between high and low-mobilities. The modes and skills to access to certain types of mobilities give to citizens different statuses. Some are esteemed as privileged citizens because they are global and financially allowed to touring, whereas others are subject to poverty and immobility. In the hyper-mobile world, cars, and mobility have emptied the real geography producing a broader sentiment of omnipotence and control (Lash and Urry, 1993). The exchange of goods and humans (through tourism consumption) engendered an emptied space where the social ties are commoditized, packaged, and sold. This process creates a sentiment of vulnerability which is addressed by the net of experts. Lash and Urry explore the role played by tour operators as the net of professionals who advise tourists (clients) respecting what areas should or not be visited. The travel agency not only absorbs the risks through the payment of capital but also imposes the ways landscapes should be contemplated. Urry, finally, is correct when he argues that some tourist destinations are preferably advertised to be sold many first-world segments while some other destinations are simply covered. The mouth-to-mouth recommendations have set the pace to a new figure: *the advertising.*

To cut a long story short, the tourist-gaze is based on three salient aspects: *the re-enchantment of consumption, time-space dimension, and visual of performing arts.* Therefore, travels re-symbolize spaces at the time they are transformed in commodities. To wit, consumers are bombarded by visual advertising and other visual stimuli to interpret these landscapes in a specific way, but in a decentralized way in comparison with other times. Consumers accept the esthetic logic of the market conferring a sacred-aura to the produced goods. The economy of signs functions conjoined to the economies of desires, which are visually stimulated by the tourism industry. Within the sociology of tourism, Urry was not only an important personality but also tried to integrate mobilities theory into a coherent conceptual paradigm to understand global geography. Urry's contribution would be impregnated in another well-known British sociologist: Kevin Meethan. In Meethan's works, tourism, and consumption should be defined seriously as something else and more complex than a pleasure-maximization process. In Meethan's development, instead of Urry or MacCannell, the concept of fluidity should be in mind. He devotes resources and time to understand the sense of place and how its borders are transformed by global tourism and the cultural industries. Like Urry, Meethan acknowledges that the sense of place is not imposed from the top down but negotiated by the involving agency. Through consumption, the self remains open to new experiences while keeping its own identity. It is important not to lose the sight of the fact that Meethan shares with Urry a concern for a cultural matrix which determines the tourist gaze, but he originally holds the thesis that tourism consumption is materialized through the articulation of substantial changes which take place in the territory. There is a type of geographic transformation where places, peoples, and cultures are constantly negotiated. Far from being an instrument of control, tourism does not alienate workers or citizens. Rather, the system exchanges, images, cultures, and people into a coherent *imagined landscape.* The idea of co-production, as well as personal interpretation, distinguishes Meethan from MacCannell and Urry. Since humans not only inhabit in the soil but they interact with others embodying expectances in forms of relations, they produce a type of explanation about who they are. Though conducive to the reproduction of *narratives of self,* tourism is enrooted in a specific history and territory which is elaborated from generation to generation. The experienced tourists feel, after consuming, entails *"active engagement in the memory work"* (Meethan, 2006, p. 9).

Tourists are not naïve or passive agents who look to maximize their pleasure but they behave in the mid of fluid and cyclical context. Starting from the premise that *hybridity* recreates complex scenarios where the agent is unfamiliar with their inner-world, as Meethan suggests, it is almost impossible to understand tourism through the cultural matrix (that Urry mentions). This point critically separates Meethan from Urry, embracing the Foucauldian thesis of a *genealogy of tourism* (Meethan, 2003, 2004, 2006, 2014). We are psychologically moved to get countless experiences but this does not appear to be the main reason for tourism. We need to travel to be in contact with the Otherness. In this host-guest interaction is crucial to understand tourism. The sense of place seems not to be the place where we live, but the place we simply imagine. This is the reason why any movement of Diaspora requires of an imagined (if not fabricated) past. For Meethan, tourism is more than an act of hedonist consumption it represents a philosophical quest of the alterity to rediscover the self. As a rite of passage where the ideals of normalcy and exoticness are blurred, tourism starts a trip of self-discovery where the liminal factor occupies a central position (Meethan, 2004). This point will be expanded in the next chapter. Basically, because of time and spatial issues, it was almost impossible to review the works of all sociologists who have studied tourism. Instead, we place the works of three senior sociologists in the lens of scrutiny, Dean MacCannell, John Urry and Kevin Meethan. From different and convergent angles, each one has elicited his own definition of tourism which is confronted with the economic-based paradigm. His legacy and main theories were continued by other sociologists over the decades. The problem of their conceptions, as explained in the first chapter, was associated with the belief that tourism emerged just after the industrial revolution, as well as the pejorative definition they develop from tourism. In this direction, a third way is strongly necessary. In consonance with the founding parents of sociology and anthropology who envisaged the advance of market and industrialism, as well as modern tourism as an objectifying process, Urry, MacCannell or even Meethan enthralled as the exponents of the sociology of tourism but paradoxically closing the doors for those who have evidence that tourism can be very well traced back beyond the borders of Thomas Cook or the capitalist system.

2.4   CONCLUSION

As discussed, the present conceptual chapter introduced readers to the epistemology of tourism fields. It centered the analysis on the dichotomies between the sociology of tourism which headed a critical debate on tourism and the economic-based paradigm more inclined to protect the tourist destination. While the former signaled to the social nature of tourism, the latter emphasized on the profitability of the tourist system. Based on the tourist and its psychological preferences as the main object of study, tourism management resoundingly failed to offer an all-encompassing model to help to understand what tourism is. What is equally important, even if the sociology of tourism launched to get an all-encompassing model to define tourism, they started from an old prejudice which was widely debated in the introductory chapters that form this book. Tourism was considered as an industrial phenomenon limited to the misconception of tourism scholars in the inception of the discipline. For that reason, the current chapter triggers a hot debate revolving around the nature of tourism, which needs to be continued. In the next chapters, we shall define tourism as a rite of passage, explaining why other forms of tourism can be easily found in ancient cultures.

KEYWORDS

- **ancient civilization**
- **consumption**
- **frustration**
- **leisure**
- **revitalization**

REFERENCES

Airey, D., & Tribe, J., (2006). *An International Handbook of Tourism Education*. Routledge, London.

Amoah, V. A., & Baum, T., (1997). Tourism education: Policy versus practice. *International Journal of Contemporary Hospitality Management, 9*(1), 5–12.

Aramberri, J., (2001). The host should get lost: Paradigms in the tourism theory. *Annals of Tourism Research, 28*(3), 738–761.

Archer, B., (1991). The value of multipliers and their policy implications. In: Medlik, S., (ed.), *Managing Tourism* (pp. 15–32). Butterworth Heinemann, Oxford.

Augé, M., (1995). *Non-Lieux*. Verso, London.

Buhalis, D., & Costa, C., (2006). *Tourism Management Dynamics: Trends, Management and Tools*. Routledge, Abingdon.

Butler, R. W., (1991). West Edmonton mall as a tourist attraction. *Canadian Geographer/ Le Géographe Canadien, 35*(3), 287–295.

Chen, J. S., & Hsu, C. H., (2000). Measurement of Korean tourists' perceived images of overseas destinations. *Journal of Travel Research, 38*(4), 411–416.

Cohen, E., (1972). Towards a sociology of international research. *Social Research, 39*(1), 164–182.

Cohen, E., (1974). Who is a tourist?: A conceptual clarification. *The Sociological Review, 22*(4), 527–555.

Cohen, E., (1988). Authenticity and commoditization in tourism. *Annals of Tourism Research, 15*(3), 371–386.

Craik, J., (2002). The culture of tourism. In: Rojek, C., & Urry, J., (eds.), *Touring Cultures* (pp. 123–146). Routledge, Abingdon.

Crompton, J. L., (1979). An assessment of the image of Mexico as a vacation destination and the influence of geographical location upon that image. *Journal of Travel Research, 17*(4), 18–23.

Dann, G., & Cohen, E., (1991). Sociology and tourism. *Annals of Tourism Research, 18*(1), 155–169.

Dann, G., (2002). *The Tourist as a Metaphor of the Social World*. CABI, Wallingford.

Durkheim, E., (1976). *The Elementary Forms of the Religious Life*. Abingdon: Routledge.

Durkheim, E., (2014). *The Division of Labor in Society*. New York: Simon and Schuster.

Fernandez, F. L., (1967). *Teoría y Técnica del turismo [Theory and Practice of Tourism] Tomo I*. Editorial Nacional, Madrid.

Fodness, D., (1994). Measuring tourist motivation. *Annals of Tourism Research, 21*(3), 555–581.

Frechtling, D., (1991). Key issues in the US travel industry futures. In: Medlik, S., (ed.), *Managing Tourism* (pp. 82–93). Butterworth Heinemann, Oxford.

Gnoth, J., (1997). Tourism motivation and expectation formation. *Annals of Tourism Research, 24*(2), 283–304.

Hall, C. M., (2013). Policy learning and policy failure in sustainable tourism governance: From first-and second-order to third-order change? In: *Tourism Governance* (pp. 249–272). Routledge, Abingdon.

Houlot, A., (1961). Le Turisme et La Biblie. Revue l'Académie Internationale du Turisme. Monaco.

Hoxter, A. L., & Lester, D., (1988). Tourist behavior and personality. *Personality and Individual Differences, 9*(1), 177–178.

Iso-Ahola, S. E., (1982). Toward a social psychological theory of tourism motivation: A rejoinder. *Annals of Tourism Research, 9*(2), 256–262.

Jenkins, C. L., (1991). Tourism Policies in developing countries. In: Medlik, S., (ed.), *Managing Tourism* (pp. 269–277). Butterworth Heinemann, Oxford.

Jennings, G., (2001). *Tourism Research.* John Wiley and sons Australia, Ltd.

Jeong, C., (2014). Marine tourist motivations comparing push and pull factors. *Journal of Quality Assurance in Hospitality & Tourism, 15*(3), 294–309.

Jiménez, G., & Luis, F., (1986). *Teoría Turística: Un Enfoque Integral Del Hecho Social [Theory of Tourism: An All Encompassing Approach].* Universidad Externado de Colombia, Bogotá.

Korstanje, M. E., (2007). The origin and meaning of tourism: Etymological study. *E-Review of Tourism Research, 5*(5), 100–108.

Korstanje, M. E., (2018). Towards new paradigms in tourism fields: An anthropological perspective. *International Journal of Tourism Anthropology, 6*(2), 108–112.

Laws, E., & Scott, N., (2015). Tourism research: Building from other disciplines. *Tourism Recreation Research, 40*(1), 48–58.

Lee, C. C., & Chang, C. P., (2008). Tourism development and economic growth: A closer look at panels. *Tourism Management, 29*(1), 180–192.

Lee, G., O'Leary, J. T., Lee, S. H., & Morrison, A., (2002). Comparison and contrast of push and pull motivational effects on trip behavior: An application of a multinomial logistic regression model. *Tourism Analysis, 7*(2), 89–104.

Leiper, N., (1979). The framework of tourism: Towards a definition of tourism, tourist, and the tourist industry. *Annals of Tourism Research, 6*(4), 390–407.

Leiper, N., (1983). An etymology of "tourism". *Annals of Tourism Research, 10*, 277–281.

Lickorish, L. J., & Jenkins, C. L., (2007). *Introduction to Tourism.* Routledge, Abingdon.

MacCannell, D., (1973). Staged authenticity: Arrangements of social space in tourist settings. *American Journal of Sociology, 79*(3), 589–603.

MacCannell, D., (1976). *The Tourist: A New Theory of the Leisure Class.* University of California Press, Berkeley.

MacCannell, D., (1984). Reconstructed ethnicity tourism and cultural identity in third world communities. *Annals of Tourism Research, 11*, 375–391.

MacCannell, D., (1988). Turismo e identidad [tourism & identity] (Juncar ed.). Madrid. In: Todorov, T., & MacCannell, D., (1992) (eds.), *Empty Meeting Grounds: The Tourist Papers.* London: Routledge.

MacCannell, D., (2001). Tourist agency. *Tourist Studies, 1*, 23–37.

MacCannell, D., (2002). *Empty Meeting Grounds: The Tourist Papers.* Routledge, London.

MacCannell, D., (2011). *The Ethics of Sightseeing.* Berkeley, CA: University of California Press.

MacCannell, D., (2012). On the ethical stake in tourism research. *Tourism Geographies, 14*, 183–194.

Mannell, R. C., & Iso-Ahola, S. E., (1987). Psychological nature of leisure and tourism experience. *Annals of Tourism Research, 14*(3), 314–331.

McCabe, S., (2001). The problem of motivation in understanding the demand for leisure day visits. *Journal of Travel & Tourism Marketing, 10*(1), 107–113.

McNaughton, D., (2006). The "host" as uninvited "guest": Hospitality, violence and tourism. *Annals of Tourism Research, 33*(3), 645–665.

Medlik, S., (1991). *Managing Tourism.* Butterworth Heinemann, Oxford.

Meethan, K., (2003). Mobile cultures?, hybridity, tourism and cultural change. *Journal of Tourism and Cultural Change, 1*(1), 11–28.

Meethan, K., (2004). To stand in the shores of my ancestors. In: Coles, T., & Timothy, D., (eds.), *Tourism, Disaporas and Space* (pp. 139–150). London: Routledge.

Meethan, K., (2006). Introduction: Narratives of place and self. In: Meethan, K., Anderson, A., & Miles, S., (eds.), *Tourism Consumption and Representation* (pp. 1–23). Wellingford: CABI.

Meethan, K., (2014). Mobilities, ethnicities and tourism. In: Lew, A., Hall, M. C., & Williams, A., (eds.), *Tourism* (pp. 240–250). Wiley Blackwell, New York: NY.

Meethan, K., Anderson, A., & Miles, S., (2006). *Tourism, Consumption and Representation: Narratives of Place and Self.* CABI, Wallingford.

Moutinho, L., & Vargas-Sanchez, A., (2018). *Strategic Management in Tourism, CABI Tourism Texts.* Cabi, Wallinford.

Munné, F., & Codina, N., (2002). Ocio y tiempo libre: Consideraciones desde una perspectiva psicosocial [Leisure and Freetime: Considerations revolving around a psychosocial perspective]. *Revista Licere-Brasil, 5*(1), 59–72.

Okumus, F., & Yagci, O., (2006). Tourism higher education in Turkey. *Journal of Teaching in Travel & Tourism, 5*(1, 2), 89–116.

Page, S., (2007). *Tourism Management: Managing for Change.* Routledge, Abingdon.

Pearce, D. G., & Butler, R. W., (1993). *Tourism Research: Critiques and Challenges.* Taylor & Francis, London.

Pearce, D. G., (1999). Tourism in Paris studies at the microscale. *Annals of Tourism Research, 26*(1), 77–97.

Pearce, D. G., (2012). *Frameworks for Tourism Research.* Cabi, Wallingford.

Pearce, P. L., & Lee, U. I., (2005). Developing the travel career approach to tourist motivation. *Journal of Travel Research, 43*(3), 226–237.

Pearce, P. L., (1982). Perceived changes in holiday destinations. *Annals of Tourism Research, 9*(2), 145–164.

Pearce, P. L., (1985). A systematic comparison of travel-related roles. *Human Relations, 38*(11), 1001–1011.

Pearce, P. L., (1991). Introduction: The tourism psychology conversation. *Australian Psychologist, 26*(3), 145–146.

Plog, S., (1973). Why destination areas rise and fall in popularity. *The Cornell Hotel and Restaurant Administration Quarterly, 13*(3), 13–16.

Prayag, G., Hosany, S., & Odeh, K., (2013). The role of tourists' emotional experiences and satisfaction in understanding behavioral intentions. *Journal of Destination Marketing & Management, 2*(2), 118–127.

Raymond, E. M., & Hall, C. M., (2008). The development of cross-cultural (mis) understanding through volunteer tourism. *Journal of Sustainable Tourism, 16*(5), 530–543.

Reynolds, T., (1965). The Roles of consumer in image building. *California Management Review* (pp. 69–76). Spring.

Rice, J., & Khanin, D., (2019). Why do they keep coming back? The effect of push motives vs. Pull motives, and attribute satisfaction on repeat visitation of tourist destinations. *Journal of Quality Assurance in Hospitality & Tourism, 20*(4), 445–469.

Richards, G., (2018). Cultural tourism: A review of recent research and trends. *Journal of Hospitality and Tourism Management, 36*, 12–21.

Richter, L., & Waugh, W. (1995). Terrorism and tourism in logical companions. In: Medlik, S., (ed.), *Managing Tourism* (pp. 318–326). Butterworth Heinemann, Oxford.

Salazar, N. B., & Graburn, N. H., (2014). *Tourism Imaginaries: Anthropological Approaches*. Berghahn books, Oxford.

Smith, V. L., (2012). *Hosts and Guests: The Anthropology of Tourism*. University of Pennsylvania Press: PE, Philadelphia.

Snepenger, D. J., (1987). Segmenting the vacation market by novelty-seeking role. *Journal of Travel Research, 26*(2), 8–14.

Theobald, W. F., (2005). The meaning, scope, and measurement of travel and tourism. *Global Tourism, 3*, 23–48.

Tribe, J., (1997). The indiscipline of tourism. *Annals of Tourism Research, 24*(3), 638–657.

Tribe, J., (2000). Indisciplined and unsubstantiated. *Annals of Tourism Research, 27*(3), 809–813.

Tribe, J., (2010). Tribes, territories and networks in the tourism academy. *Annals of Tourism Research, 37*(1), 7–33.

Tucker, H., (2003). The host-guest relationship and its implications in rural tourism. In: Hall, D., Roberts, L., & Mitchel, M., (eds.), *New Directions in Rural Tourism* (pp. 80–89). Ashgate, London.

Um, S., & Crompton, J. L., (1990). Attitude determinants in tourism destination choice. *Annals of Tourism Research, 17*(3), 432–448.

Urry, J., (1990). The consumption of tourism. *Sociology, 24*(1), 23–35.

Urry, J., (2002). *The Tourist Gaze*. Sage, London.

Urry, J., (2007). *Introduction. Viajes y Geografías*. Prometeo, Buenos Aires.

Uysal, M., & Crompton, J. L., (1985). An overview of approaches used to forecast tourism demand. *Journal of Travel Research, 23*(4), 7–15.

Uysal, M., & Jurowski, C., (1994). Testing the push and pull factors. *Annals of Tourism Research, 21*(4), 844–846.

Van, D. J., (1991). Can future research contributes to tourism policy? In: Medlik, S., (ed.), *Managing Tourism* (pp. 3–14). Butterworth Heinemann, Oxford.

Virilio, P., (2006). *Speed and Politics: An Essay on Dromology*. Polizzotti, Mark (trad.).: Semiotext (e), Los Angeles, CA.

Weber, M., (2009). *From Max Weber: Essays in Sociology*. Routledge, London.

Yiannakis, A., & Gibson, H., (1992). Roles tourists play. *Annals of Tourism Research, 19*(2), 287–303.

CHAPTER 3

Tourism as a Rite of Passage

ABSTRACT

The earlier two chapters of the present editorial project served to set a critical position regarding what scholars dubbed as "economic-based paradigm" and its applications in the constellations of tourism. This theory not only punctuates to tourism as a sub-sector of the service industry but also pays excessive attention to its economic impact in the community. These chapters reviewed the limitations and conceptual inconsistencies of the economic-centered theory. Complementary, we now dare to present our own definition of tourism, which so to speak is considered a rite of passage. Although historians reached some consensus to indicate tourism emerged as a result of combined factors like the reduction working hours and the technological breakthrough no less true seems to be that anthropology advanced with leaps and bounds in identifying ancient forms of tourism, probably in ancient civilizations like Romans, Babylonians, and Assyrians. With different names and shapes, tourism is present in the history of mankind and sedentary tribes from immemorial times.

3.1 INTRODUCTION

In the earlier chapters, we not only explored the contributions and limitations of different scholars to tourism fields but also brought some reflections on the nature of leisure and tourism, above all, from the lens of management, sociology, and cultural studies. This chapter is reserved to offer a new alternative interpretation in view of the methodological problems the discipline faces these days. And of course, in so doing, we shall answer to the question: what is tourism?

Although the content of the chapter centers on themes we are not dealing with the moment, in view of the current epistemological crisis of the discipline, the point deserves our efforts. Having said this, let us add that the current chapter synthesizes my own experience as writer, researcher, editor, and of course fieldworker while paves the ways for an emerging conceptual platform that helps understanding tourism *as a rite of passage* within the fields of leisure (dream-like theory). But and here is the point, a passage to where?

Decades earlier, John Tribe alerted that tourism research was experiencing an epistemological fragmentation which invariably ushered professional researchers into a puzzling situation. The lack of interest or the indifference of an Academia, which does not dialog with researchers, associated to some flexible methodological approaches to define tourism accelerated a state of chaos which resulted in an irreversible knowledge fragmentation, Tribe sadly concludes. Different academic schools-as we have reviewed in the former chapters-have developed and monopolized their own definition of tourism. In fact, what tourism is or how it evolved in the time remains as an object of a dispute to date. In his seminal text, which is entitled the *scientification of tourism,* Jafar Jafari called the attention to the rise of a new knowledge-based platform in tourism fields, where any individual interpretation would set the pace to more objective evaluations not only based on empiricism but in blazing the trail of other tourism-related researchers in the years to come. As Thirkettle and Korstanje (2013) put it, the struggle for emergent schools, which struggled to monopolize and impose their own interpretations, prompted a much deeper dispersion almost impossible to control. Instead of coordinating efforts to forge a more efficient and harmonized method, tourism-related scholars adopted transdisciplinarity as a vehicle towards scientific maturation. From its onset, applied-research in tourism has been influenced by a business-centered paradigm in which case, tourism was naively defined as an industry in lieu of an ancient social institution. Rather than achieving the desired results, studies focused on the needs of finding new segments (demand) to satisfy the needs of suppliers.

As above noted, the economic-based paradigm not only focused on the tourist destination leaving other themes without or with marginal attention but also emphasized on the economic benefits of tourism, leading others to think in tourism as a growing industry (Page and Connell, 2006). In this way, tourism management was imagined as an instrument for governments and private organizations to organize territories with more efficiency (political governance). Over the years, such a belief, which systematically accompanied tourism research, as well as the different publications, leaves researchers to understand tourism as a millenarian or mythical institution (Wearing and Wearing, 2001; Higgins-Desbiolles, 2006). As Professor Van Doorn (1993) observes, professional fieldworkers have historically coordinated efforts to eliminate all risks which may place the industry in jeopardy, departing from the premise that the destination image and its attractiveness as the only key factors of their respective works. Hence, quantitative, and statistical methods were certainly prioritized in view of the variables that needed to be measured. Focusing on the needs of measuring, instead of understanding, the economic-centered paradigm adopted a profit-oriented strategy, excluding other voices from the top-ranked journals (Thirkettle and Korstanje, 2013). Although Tribe never was more accurate on the point, the knowledge fragmentation has been increased simply because the discipline retained a profit-oriented goal, adjusted to the demand. Tourism as an object of study went through its own dispersion in many forms, and what is more important each one adaptable to a specific international demand. To set a clear example, in the 80s and 90s decades, the academicians hold to sustainable tourism to denote a type of new tourism linked to sustainability or to segments which looks not to contaminate the planet. Dark tourism, today, surfaced as the leading topic which is understood as the new segment of tourists who need to visit spaces of mass death or disaster, and so forth. This suggests that the different definitions revolving around tourism are not given by the nature of the object, but by the needs of segmenting the demand. The more the segments, the more the forms (definitions) of tourism. Having said this, heritage, dark, and sustainable tourism are versions of the same issue. Epistemologically speaking, what tourist feels or believes situated as the primary criterion for field-working probably ignoring that sometimes interviewees are unfamiliar of their emotions or inner-world or simply lie to protect their interests. Therefore, this chapter attempts to redirect this hot debate confronting not only to already-established literature but also giving a new fresh alternative to understand tourism as a rite of passage, which is based in the subject as myth-producer. Particularly one of the limitations of the discipline nowadays seems to be the excessive credibility given to interviews, questionnaires, and other obtrusive methods.

3.2   THE RITE OF PASSAGE

Returning to the nature of tourism, it is very hard to decipher almost the main theories within social sciences. Because of some constraints in time and space, we were forced to exclude some seminal texts from this discussion. In this section, we held the thesis tourism is *a rite of passage* whose original function consists of engendering a territorial and identity's dislocation. Such a rupture is activated in order for the people to pretend themselves in other roles. Similarly, to the function of dreams (as a play in terms of Cohen), tourism revitalizes all material deprivations balancing the personal frustrations-if not situating the self into a newly renovated status. The break with routine or day-to-day rules is almost a universal practice in all cultures, as professor Bronislaw Malinowski (1994) eloquently said. In his studies of exchange in Melanesia, Malinowski contends that biological instincts in humans such as metabolism, reproduction, safety, movement, bodily comfort, growth, and health can be traced and observed in many aboriginal communities. Each one contains a cultural response which is crystallized as a social institution. For example, reproduction is for kinship what bodily-comfort is for shelter-games. In perspective, he recognized that the needs of recreation and playful rest are vital for the culture. Building their own forms of the escapement, cultures do the correct thing by gaining further adaptation to the environment. Those communities which fail for developing the necessary instrument for recreation are doomed to be wiped out (Malinowski, 1944). However, our polish anthropologist does not deepen his analysis in regards to the recreational behavior in aboriginal tribes. Nowadays, anthropology has produced a substantial background in order for lay-readers to understand what a rite of passage means. As a part of the world of rituals, it signals to a celebration that happens whenever a member of the community leaves a former group to be introduced in another new one, and of course, in a different status. In his book, *Les rites de Passage,* Van Gennep argues that each community has its own rites of passages where peoples, roles, and gifts are continuously exchanged. One of the aspects that characterizes these liminal rituals seems to be the needs of *physical displacement*, which sometimes places the candidate in temporal isolation to be reintroduced at a later day in the new group. These isolations are based on two types of separations, the distinction between males and females, and profane and the sacred (Van Gennep, 2011). In view of this, rites of passage comprise three facets which are, separation, liminality, and incorporation. The efficiency of rituals corresponds with its capacity to detach a person from the in-group rules. In the induced transition, threshold candidates avoid any direct contact with their former groups at least until the rite is completed. Not only candidates should demonstrate their worthies and virtues, but skills to be esteemed as a free man. One might think that whenever tourists avoid direct contact with other conational tourists or are prone to diversity to learn from far-away cultures, we are in the presence of a rite of passage. Echoing Van Gennep, Victor Turner offered an interesting conceptual model to understand the connection between passage and liminality. His study was certainly centered on an African tribe, Ndembu (Zambia), a case which materialized in his Doctoral thesis completion. In consonance with Van Gennep, Turner points out that the rite of passage should be distinguished in three stages, *pre-liminal phase (separation), liminal phase (transition)* and *post-liminal phase (reincorporation)*. The role of liminality is crucial to determine the new status of candidates, Turner concludes. This happens simply because the transitional stage or in-between states are tangible mechanisms adopted by the community to tolerate ambiguity. Inscribed in a limbo, candidates are tested to overcome numerous obstacles and dangers while achieving a much deeper sentiment of communitas (Turner, 1995). Not surprisingly, rites of passage are not limited to aborigines or tribal organizations; rather there are clear examples in West as baptism, tourism, graduate trips, Christmas, or New Year celebration among many others. In earlier works, Korstanje, and Busby noted that:

> *"We can conclude that renovation of norms that entails the return is enrooted in the figures of baptism, guilt, sacrifice, and expiation. This moral process can be compared with social duties or rules visitors abide by every day. These forces not only determine individual behavior but also pave the pathways towards a new reinsertion. This eternal return to day-to-day life (once the vacation is over) demonstrates an ambivalent nature. On the one hand, we change in some way but certainly it is unquestionable we are subject of a process of forgiveness. On another hand, there is continuity because we were introduced in the same real before our departure"* (Korstanje and Busby, 2010, p. 107).

This citation shows the real meaning and evolution of tourism as a social institution and of course comparable to hospitality. What is worth of mention seems to be inferring to what extent, the inversion of rules is a key factor that leads liminal candidate (tourists) to a new status. To put this in bluntly, those sedentary societies (following the example of Westerners) where the labor (sacrifice) is the rule that dictates civilized coexistence, the maximization of pleasure becomes the necessary rite of passage. Otherwise, archaeology, and history witnessed how aborigines (in times where Spanish colonizers have not arrived at Americas) who were unfamiliar with the logic of work, since they worked only for subsistence, rites of passages were marked by pain, tests, and games of forces. European empires as Spain and Portugal, once introduced in the continent, expanded their hegemony by the exploitation of aborigines. They introduced the needs of work in societies whose economies were not based on surplus. Most certainly, the rites of passage of these cultures experienced a radical shift which prompted their cultural death. The western model of work-leisure (as MacCannell puts it) was imposed to aboriginal tribes in a way that subordinated their culture into the westerner-gaze (Urry). From that moment onwards, not only the colonization of this different "Other" was consolidated but the colonial discourse of recreational travel based on the European legal jurisprudence mushroomed. The function of law, hospitality, and recreation as they were coined in Europe, were forcedly adopted by the fourth world. In so doing, it contributed to the expansion of Christianity in the Americas. Then, what is the mission of tourism-researchers? Epistemologists, experts, and pundits should understand that the current modern tourism (which may be dubbed Anglo-tourism) not only seems to be defined as a rite of passage but also other forms of recreational travel should be explored. Levi-Strauss and structuralist method envisaged the creation of a periodic table of cultures, researchers should find a periodic table of tourism(s). This was originally what MacCannell, Urry or Krippendorf dreamed but failed to perform. The division between totems vs. tourism is not enough to explain the nature of the phenomenon.

It is unfortunate that the historians of tourism never looked beyond the walls of the Middle Age, considering erroneously that tourism emerged just after the industrial revolution. Different forms of tourism have been present in ancient empires such as Babylonians, Assyrians, and even Romans. To a closer look, the term *ferias* (which comes from Latin) was originally reserved as a leave assigned to those roman citizens who have worked one year for the empire. This leave, which was valid for 3 months, facilitated the visit to relatives in the periphery. In this way, the Roman Empire affirmed its control and solidarity with the periphery, enhancing the social bond. Etymologically speaking, the term holiday in modern languages such as Portuguese and German stems from this ancient leave: Das Ferias (Por) and Die Ferien (Ger). For some reason, which was associated with the lack of interests in ancient history, tourism-related historians theorized on the Grand-Tour as the only precondition for a type of pre-tourism era. Of course, in the Middle Age, the degree of mobility was notably reduced because of the feudal wars, the attacks of bandits, or simply the poor-quality life. However, this does not mean that tourism was unknown for Europeans or Asians (Korstanje, 2009). Here, anyway, two assumptions should be done. If we have robust evidence that the term tourism was originally used since 18th-century A.C. (or near circa), why do we suppose another thing than tourism-as an industry-is indeed a modern invention? secondly, may we study two topics dotted with different names as the same object?

On the one hand, the themes are always framed in the capacity of language to seize (name) them. As social researchers, we are hand-tied to approach a topic which was no previous presence in our language. As a social construal, the research theme is previously defined according to an earlier paradigm which precedes us. It is common to see how different themes analogously speak from the same object and vice-versa we use the same name for two different objects of study. Through the analogy, researchers may investigate different objects as only one. This seems to be a clear example between holidays and tourism. Are we talking about the same issue?

On another, social scientists are sometimes encapsulated into a Eurocentric ideology that leads them to think that modern institutions (like tourism) were a product of modernity. Needless to say, the modern tourism which was re-affirmed just after the WWII end, seems to result from the rise and expansion of Anglo-Saxon Empire (the UK and the US respectively) but this does not authorize to hold that tourism is a modern institution. There were ancient and non-western forms of tourism in history documented in interesting studies (Mooney, 1920; Adams and Laurence, 2012; Nimis, 2004; Harlow and Laurence, 2002).

3.3   ASKING FURTHERLY ABOUT THE NATURE OF TOURISM

The evolution of the discipline has taken place not without some discrepancies. Basically, we can gather the theorists together and classify them in two clear-cut academic groups. While some voices-certainly prioritizing the evolution of the industry-approached tourism as a genuine generator and distributor of wealth. Those authors who take part of this collective believe that capitalist and democratic societies have further opportunities to promote tourism than undemocratic ones. The fact is that tourism can be understood as a virtuous instrument planned to create employment and boost the local economy (through a fairer distribution of wealth). From this perspective, the development, and the subsequent trickle-down theory, which punctuates that tourism stimulates further foreign investments and tax reduction in which case the economy experiences a rapid recovery just after stock and market crises, has its risks. In this vein, neither development theory nor tourism seems to be successful in all contexts. Those cultures, which were traversed by a bloody past or torn in civil wars, are more prone to social maladies that impede tourism development (McKercher, 1993; Leiper, 2000; Smith, 1994). Another group, instead, holds the thesis that tourism orients as a neo-colonial force which creates certain economic dependency in underdeveloped countries. These scholars not only exert a radical criticism on the colonial legacy but warn on the rise and expansion of tourism as a new ideological form of control that continues the Global-North's hegemony. To a closer look, tourism should be evaluated and judged by its effects on the local economy. Richest tourist-delivering countries invest heavily in tourist-receiving countries (in infrastructure, transport, and hospitality) affecting not only the local economy but repatriating the surplus and capital to the same place where tourists depart from (Turner and Ash, 1975; Girault, 1978; Britton, 1982; Wellings and Crush, 1983; Leep, 2008; Chaperon and Bramwell, 2013).

The needs of displacing for *being there, a vital element that distinguishes tourism today,* can be equaled to the first ethnologists and social scientists who launched to the unknown to document Non-western cultures in the colonial period. Aside from the scientific interests of these explorations, Europe expanded the colonial order to the periphery imposing not only a cultural matrix but their products and trade (Pratt, 2007; Caton and Santos, 2007; Bandyopadhyay and Nascimento, 2010; Korstanje, 2012). It is important not to lose the sight of the fact even if with some substantial disagreements, both academic traditions give to *the economic factor* a privileged position.

The anthropologist Agustin Santana Talavera has convincingly explained that the already-existent theories in tourism fields can be organized in six great families: (a) commercial hospitality; (b) an instrument of democracy; (c) a subtype of leisure; (d) a form of cultural expression; (e) a process of acculturation; and (f) a discourse that strengthens the colonial dependency between center and periphery (Santana Talavera, 1997). Though it is hard to imagine tourism without the pay-for logic, it is important not to lose sight other theories have said something on this. To what extent may we understand that tourism and consumption are interlinked?

For readers to understand the nature of tourism it is important to review the main contributions of Jost Krippendorf who does not need the previous presentation. The Swiss-born economist, Jost Krippendorf acknowledges that tourism serves as a mechanism of revitalization which prevents conflicts while rechanneling the social change. As something more complex than a mere industry, tourism is essentially a social institution that transversally transcends modernity or the industrial revolution. Per his viewpoint, at the time people travel, what they look seems to be associated with the basic needs of resting. The psychological frustrations that happened during the work are revitalized and sublimated through the introduction of the escapement. As a sacred space where pain momentarily disappears, tourism denotes some reversibility in the rules of productivity. In our holidays, we play to be what we wish to be, but we are not in real life. This is the reason why, as Krippendorf adheres, leisure, and tourism are inextricably intertwined. Tourism enacts not only an ideological message to citizens in order for them to be proud of belonging to a privileged society, but also exhibits the foundational cultural values of society. For that, each society develops different forms of leisure and tourism. Combining anthropological insights with their own studies in economy, Krippendorf leaves an interesting model that helps understanding tourism from the lens of sociological theory. He remarks that the origin of tourism should be traced back to the inception of the sedentary phase in human history. To wit, the decline of happiness within western societies lay-people today experience derives from the highly-degree of alienation suffered in day-to-day life. At once the productive system is more oppressive; further leisure is needed to counter-balance the created material asymmetries (Krippendorf, 1975, 1982, 1986, 1987, 1989, 1995).

In this respect, it is important to mention that tourism behavior rests on what anthropologists have studied as *the metaphor of lost-paradise.* Countless mythologies and religions evince the needs of humans to return to an exemplary center which-for many reasons-was simply lost. The eternal quest for this exemplary center corresponds with the attachment with the mother's womb. This top-down cosmology gives, as a result, a hierarchy of exploiting and exploited classes. In any societal order, the elite monopolizes not only the means of production, but also the allegories by which the work-force is subordinated, or in terms of MacCannell alienated.

Erik Cohen takes some distance from the works of Urry and MacCannell which were debated in Chapter 2. For him, the nature of tourism cannot be explained through the figure of escapement only. Neither industrialism nor the rise of an *aesthetized* modernity was sufficient conditions to explain the origin of tourism. Although Cohen agrees with Urry and MacCannell that tourism is a modern institution, he defines tourism *as a type of commercialized hospitality* where there is a temporal dissociation (liminoid space) of self and the societal rules. In view of this, tourism upends not only the logic of work, but also the societal rules leading the self towards enhancement and a temporal state of pleasure-maximization. In Cohen's account, the concept of *play* is essential in the difficult task of defining tourism (Cohen, 1972, 1979, 1988).

Another exponent of the field, who deserves further attention, is doubtless Valene Smith. In her stance, the social function of tourism seems to be the revitalization of social scaffolding. However, since *"as work gives way to leisured mobility, individuals find re-creation in a variety of new contexts. Different forms of tourism can be defined in terms of the kind of leisure mobility undertaken by the tourist"* (Smith, 2012, p. 5). At this stage, Smith explored the effects of the activity over the local community as well as the problems the tourist bubble generated. The *guests-and-hosts encounter* may be very well a problematic issue if policymakers do not regulate the economic asymmetries created by the industry. The concept of acculturation is one of the most interesting points, placed by Smith in her original texts. However, unlike MacCannell or Krippendorf, she does not provide a thorough explanation of what tourism is.

In consonance with this, Graham Dann coins the term *phantasy world* to materialize the dynamic of tourism. In the world which is characterized by the anomie, the ego runs serious risks of fragmentation (dissociated self). To avoid this, the ego should be culturally protected. This *phantasy world* not only protects but also enhances the frustrated ego. Dann goes on to say that tourism should be defined as a dominant metaphor, an allegory of an ever-changing world. Behind the figure of tourist lies the sociocultural background that forged the cultural values of society and its potential evolution. To put this slightly in other terms, tourism corresponds with a change of environment, which only is feasible by displacement. In this quest for novelty, or authenticity, ethnography offers a good opportunity to find answers that clarify *the meaning of tourism* (Dann, 1977, 1996, 2002). Still further, Nelson Graburn argues convincingly that tourism should be comparable as a rite of passage, a type of sacred-journal the re-founds the societal structure. The meaning of tourism associates to the play where everything can happen. This degree of contingency opens the doors to social conflict and unforeseen reactions between hosts and guests. We do things that usually we do not do at home because magic occupies a central position in tourism behavior (Graburn, 1976, 1983, 1989).

> *"The food and drink might be identical to that normally eaten indoors, but the magic comes from the movement and the non-ordinary setting. Furthermore, it is not merely a matter of money that separates the stay-at-home from extensive travelers. Many very wealthy people never become tourists, and most youthful travelers are, by western standard, quite poor"* (Graburn, 1976, p. 24).

In sum, Graburn takes his conceptual model from the previously discussed insights formulated by MacCannell (totem), Cohen (lost paradise) and Smith (host-guest encounter). Lastly, Noel Salazar (2015) holds that the real connection with the alterity is given by a fabricated experience, where this Other is externally designed and consumed. Tourists often are in quest of familiarity in an unfamiliar scenario. Locals who are part of the attraction are stereotyped according to a much deeper ideological discourse formed in a society where tourists come from. While touring, these transnational imaginaries align to the personal experience. For Salazar tourism and consumption are two sides of the same coin.

3.4   THE LIKE-DREAM THEORY

At some perspective, tourism in practice is an inherent part of the subsystem of leisure (which is associated with other institutions and activities such as reading, films, sporting spectacles, and the theater, among others). One of the most important functions of leisure is to maintain a balance in the social system. At a first glimpse, leisure is the foundation not only of work but also of the socialization of individuals in the different cultural aspects and values which are important for society. Initially, authors such as Freud and Jung defined the dream-like state as a psychical and biological regulatory mechanism, associated with sleep. For both authors, dreams are an activity of the unconscious produced by the libido whose principal characteristic is to compensate the psychic system for the different frustrations experienced while awake (Freud, 1961). By fulfilling the repressed desire, dreams combine and articulate levels of thought into one coherent whole. However, for Jung, dreams should be de-codified in a message concerning our own 'self-knowledge.' In contrast to Freud, Jung (2013) maintains that dreams should be understood as rather more than a mere result of the repression of fear and desire, but function under the principle of fantasy evoking 'truths' about which the subject is unaware or has not registered while awake. Does this mean that tourism is an industry of fantasy? Tourism, whether for vacations or not, is defined by a dream-like process, which follows on from the pre-touristic phase, defined above as being concerned with hospitality. According to our perspective, the dream-like subsystem of society has two functions: the release of stress and a wider re-accommodation to a new situation. Whereas the first tends to reduce conflict by loosening the ties which unite society and thus leading the individual towards selfish behavior, the second refers to a dynamic whereby the subject re-inserts himself into a slightly different role. This role, which confers identity on the subject, follows cyclical processes. To the contrary to the position of Turner and Cohen, a vacationist does not change either his status or his role when he takes a holiday in a particular destination. The function of the dream-like (or dream-like) system is to preserve the different components of society, such as the political system and the productive system, thus avoiding dramatic social change. The dream-like system itself rests on three principal pillars. The first is scarceness, without which it cannot operate. Relationships between the actors cause situations of everyday deprivation and even symbolic frustration. To avoid a situation where the members completely abandon the group, as we see in the case of migration, the dream-like system gives back to the dreamers a 'motive' to belong to the group, and a theme which makes it worthwhile, and even necessary, to belong to the group. The second element is the extra-ordinary, which reminds us of the first heroes. The dream-like subsystem, like dreams themselves, permits the subject to do things which are prohibited in the waking state. In a film or sporting spectacle or another event, the subject experiences a type of dream-like cathartic meeting with his heroes, who is destined to mediate between men and the gods. Lastly, predestination gives to society concrete examples of activities that might put its very survival in danger. The dream-like system (conformed by scarcity, the extra-ordinary, and predestination), and by means of leisure activities as described by Huizinga (1950), redefines the limits of uncertainty so that the subject might anticipate an accident. As we will see, the dream-like subsystem does not operate alone but linked with four other subsystems, which are just as important or more important. Just as each fracture is bound to be re-adjusted, the dream-like system of regulation is determined by the conflicts in the future and past of humanity, and its hopes, frustrations, and contradictions. Society is composed of five subsystems which are mutually interconnected: the political subsystem, which accumulates and distributes power, the economic subsystem (which regulates scarcity), the mythical-religious subsystem (which tries to explain cosmic incongruences), the geographic subsystem (which maintains the identity, and the security of borders), and the c subsystem (which absorbs the tensions and conflicts generated by the other four subsystems, and forms a consensus which is disputed by no-one). Leisure is a part of the dream-like subsystem, and tourism is one of the many forms of leisure. Tourism, furthermore, generates a discourse which regulates the wish of the individual. Mobility, as a supreme cultural value of the west, is a right transmitted to children through different means of socialization from their earliest years. This same 'right' to mobility is encouraged by holidays (as sacred spaces dedicated to the practice of tourism, and whose objective is the creation of economic wealth), and by specific economic interests. The geographic subsystem is also important in the planning of tourism routes where tourism for recreation is safe (or unsafe).

Entertainment, which is an element of all voyages which alternate relaxation with moderate risk, is the basis of tourism as a total phenomenon and applicable to all cultures.

When the dream-like system cannot perform its function of regulation and balance, social change is the result. Still further, the mythical-religious subsystem (which includes all those who preserve knowledge such as priests, scientists, and journalists) needs a story to give any sense to the world and the events which happen. The process of the construction of myths is the foundation for cultural values which support society and around which are created different rituals, heroes, and cultural practices. The link between the mythical-religious subsystem and the dream-like subsystem is of great complexity. Cultural values which are considered to be unquestionable, such as rest, mobility, and the return, are essentially transmitted by myths of origin, or genesis, and may be observed in the different religious doctrines of the world. A temporary journey, therefore, requires a psychic relocation of the voyager, who thus experiences, in his fantasy, the necessity of a change of narrative and identity, and subsequent return to his situation of origin.

3.5 CONCLUSION

To here we have discussed the complexity of tourism as well as the like-dream theory, a conceptual platform that situates tourism-like a dream-as a social institution designed to revitalize the social frustration while regulating conflict. As such, tourism seems not to be the patrimony of modernity or a modern institution, as some scholars believe. There is substantial evidence in archaeology and anthropology that tourism was performed by other civilizations in the ancient past, even some Non-western cultures develop these types of practices. The present chapter reviewed the contributions and limitations of tourism management and the evolution of tourism sociology, which focused too much attention on the economic factor while offering a specific definition of tourism as a rite of passage. Echoing Professor Graburn's works, we sustain that tourism should be understood as a rite of passage that dislocates symbolically the self's identity in a liminal way. In this way, it is revitalized in the same form we are revitalizing just after dreaming or sleeping.

KEYWORDS

- **academia**
- **epistemological crisis**
- **history of tourism**
- **holidays**
- **modernity**
- **traveling**

REFERENCES

Adams, C., & Laurence, R., (2012). *Travel and Geography in the Roman Empire.* Routledge, Abingdon.

Bandyopadhyay, R., & Nascimento, K., (2010). Where fantasy becomes reality: How tourism forces made Brazil a sexual playground. *Journal of Sustainable Tourism, 18*(8), 933–949.

Britton, S. G., (1982). The political economy of tourism in the third world. *Annals of Tourism Research, 9*(3), 331–358.

Caton, K., & Santos, C. A., (2007). Heritage tourism on route 66: Deconstructing nostalgia. *Journal of Travel Research, 45*(4), 371–386.

Chaperon, S., & Bramwell, B., (2013). Dependency and agency in peripheral tourism development. *Annals of Tourism Research, 40,* 132–154.

Cohen, E., (1972). Toward a sociology of international tourism. *Social Research,* 164–182.

Cohen, E., (1979). Rethinking the sociology of tourism. *Annals of Tourism Research, 6*(1), 18–35.

Cohen, E., (1988). Authenticity and commoditization in tourism. *Annals of Tourism Research, 15*(3), 371–386.

Dann, G. M., (1977). Anomie, ego-enhancement and tourism. *Annals of Tourism Research, 4*(4), 184–194.

Dann, G. M., (1996). Tourists' images of a destination-an alternative analysis. *Journal of Travel & Tourism Marketing, 5*(1, 2), 41–55.

Dann, G., (2002). *The Tourist as a Metaphor of the Social World.* CABI, Wallingford.

Freud, S., (1961). Some additional notes on dream-interpretation as a whole. In: *The Standard Edition of the Complete Psychological Works of Sigmund Freud, Volume XIX (1923–1925): The Ego and the Id and Other Works* (pp. 123–138).

Girault, C., (1978). Tourism and dependency in Haiti. *Cahiers des Ameriques Latines, Serie Science de l'Homme,* (17), 23–56.

Graburn, N. H., (1976). *Ethnic and Tourist Arts: Cultural Expressions From the Fourth World.* University of California Press, Berkeley. CA.

Graburn, N. H., (1983). The anthropology of tourism. *Annals of Tourism Research, 10*(1), 9–33.

Graburn, N. H., (1989). Tourism: The sacred journey. *Hosts and Guests: The Anthropology of Tourism* (pp. 21–36). University of Pennsylvania Press, Philadelphia, PE.

Harlow, M., & Laurence, R., (2002). *Growing up and Growing old in Ancient Rome: A Life Course Approach.* Psychology Press, London.

Higgins-Desbiolles, F., (2006). More than an "industry": The forgotten power of tourism as a social force. *Tourism Management, 27*(6), 1192–1208.

Huizinga, J., (1950). *Homo Ludens (English Translation).* Taylor and Francis, London.

Jafari, J., (2001). The scientification of tourism. In: Smith, V. L., & Brent, M., (eds.), *Hosts and Guests Revisited: Tourism Issues of the 21st Century* (pp. 28–41). Cognizant Communications, New York, NY.

Jung, C. G., (2013). *Dream Analysis 1: Notes of the Seminar Given in 1928–30.* Routledge, London.

Korstanje, M. E., & Busby, G., (2010). Understanding the bible as the roots of physical displacement: The origin of tourism. *E-Review of Tourism Research, 8*(3), 95–111.

Korstanje, M. E., (2009). Reconsidering the roots of event management: Leisure in ancient Rome. *Event Management, 13*(3), 197–203.

Korstanje, M. E., (2012). Reconsidering cultural tourism: An anthropologist's perspective. *Journal of Heritage Tourism, 7*(2), 179–184.

Krippendorf, J., (1975). *Die Landschaftsfresser: Tourismus u. Erholungsl and Schaft.* [The Painter of Landscapes: Tourism and Leisure]. Hallwag, Bern.

Krippendorf, J., (1982). Towards new tourism policies: The importance of environmental and sociocultural factors. *Tourism Management, 3*(3), 135–148.

Krippendorf, J., (1986). The new tourist—turning point for leisure and travel. *Tourism Management, 7*(2), 131–135.

Krippendorf, J., (1987). Ecological approach to tourism marketing. *Tourism Management, 8*(2), 174–176.

Krippendorf, J., (1987). *The Holiday-Makers: Understanding the Impact of Travel and Tourism.* Butterworth-Heinemann, Oxford.

Krippendorf, J., (1989). *Fur Einen Anderen Tourimus: Probleme-Perspektiven.* [Other forms of tourism: new problems and perspectives]. Fischer-Taschenbuch Verl, Frankfurt am Main.

Krippendorf, J., (1995). *Freizeit & Tourismus: Eine Einfuhrung in Theorie und Politiks.* FIF, Bern.

Leiper, N., (1983). An etymology of "tourism". *Annals of Tourism Research, 10*(2), 277–280.

Leiper, N., (2000). An emerging discipline. *Annals of Tourism Research, 27*(3), 805–809.

Lepp, A., (2008). Tourism and dependency: An analysis of Bigodi village, Uganda. *Tourism Management, 29*(6), 1206–1214.

Malinowski, B., (1944). *A Scientific Theory of Culture, and Other Essays.* University of North Carolina Press, Chapell Hill, NC.

McKercher, B., (1993). Some fundamental truths about tourism: Understanding tourism's social and environmental impacts. *Journal of Sustainable Tourism, 1*(1), 6–16.

Mooney, W. W., (1920). *Travel Among the Ancient Romans.* RG Badger, New York, NY.

Nimis, S., (2004). Egypt in Greco-Roman history and fiction. *Alif: Journal of Comparative Poetics,* (24), 34–70.

Page, S., & Connell, J., (2006). *Tourism, a Modern Synthesis.* Thomson Learning, London.

Pratt, M. L., (2007). *Imperial Eyes: Travel Writing and Transculturation*. Routledge, New York, NY.

Salazar, N., (2015). Becoming cosmopolitan through traveling? Some anthropological reflections. *English Language and Literature, 61*(1), 51–67.

Santana-Talavera, A., (1997). *Antropología y turismo:¿ Nuevas hordas, viejas culturas? [Anthropology and Tourism: New Hordes, old Cultures?]*. Ariel, Barcelona.

Smith, S. L., (1994). The tourism product. *Annals of Tourism Research, 21*(3), 582–595.

Thirkettle, A., & Korstanje, M. E., (2013). Creating a new epistemiology for tourism and hospitality disciplines. *International Journal of Qualitative Research in Services, 1*(1), 13–34.

Tribe, J., (1997). The indiscipline of tourism. *Annals of Tourism Research, 24*(3), 638–657.

Tribe, J., (2006). The truth about tourism. *Annals of Tourism Research, 33*(2), 360–381.

Tribe, J., (2010). Tribes, territories and networks in the tourism academy. *Annals of Tourism Research, 37*(1), 7–33.

Turner, L., & Ash, J., (1975). *The Golden Hordes: International Tourism and the Pleasure Periphery*. Constable & Robinson, New York: NY.

Turner, V., (1995). *The Ritual Process: Structure and Anti-Structure*. Transaction Publishers, New York, NY.

Van, D. J. W., (1993). "Can future research contribute to tourism policy. In: Medlik, S., (ed.), *Managing Tourism* (pp. 1–3). Butterworth-Heinemann, Oxford.

Van, G. A., (2011). *The Rites of Passage*. University of Chicago Press, Chicago, IL.

Wearing, S., & Wearing, B., (2001). Conceptualizing the selves of tourism. *Leisure Studies, 20*(2), 143–159.

Wellings, P. A., & Crush, J. S., (1983). Tourism and dependency in Southern Africa: The prospects and planning of tourism in Lesotho. *Applied Geography, 3*(3), 205–223.

CHAPTER 4

The Relevance of Ethnography to Study Tourism Fields

ABSTRACT

The history of anthropology reveals two significant things. On one hand, the needs of discovering while protecting the "non-Western Other" before its final disappearance. The European rule not only launched to colonize new territories to amass further wealth but also imagined natives as children uncultivated and uneducated who need to be re-socialized. Literature and anthropology were the tug of war where natives were ideologically documented and portrayed. While ethnographers should be "there" in order to collect much as possible from the hosting culture, colonial officers employed ethnographies to understand better what they dubbed as "noble savages". As Mary Louise Pratt puts it, the imperial eyes not only were decided to colonize the Other but also created the needs to do it. While the imperial center marked the different-alterity, it avoided to be marked. From its onset, literature and Science were of paramount importance in the ideological construction of the European (colonial) Matrix. The colonial traveler sets the pace to the modern tourist –though echoing Pratt the same cultural mandate persists. In tourism fields, for some reason which is very difficult to precise here, remained indifferent from the benefits of employing ethnography. To fulfill this gap, the present chapter dealt with ethnography, its definitions and historical evolution, as well as the challenges for the years to come.

4.1 INTRODUCTION

There is some discrepancy at the time of defining ethnography in the social sciences. While some scholars applaud ethnography enthusiastically as an illustrative method of inquiry (Brewer, 2000; Dixon-Woods, 2003), other colleagues object its relevance considering as a type of art (Clifford, 1988; Gobo, 2008; Hammersley, 2018). The history of anthropology reveals two important things. On the one hand, the needs of discovering while protecting the "non-Western Other" before its final disappearance (Hymes, 2003). The European rule not only launched to colonize new territories to amass further wealth but also imagined natives as children uncultivated and uneducated who needed to be re-socialized. Literature and anthropology were the tug of war where natives were ideologically documented and portrayed. The local native was dispossessed form its history and replaced under the lens of novel literature, which was the only capable of describing him. In so doing, Europeans passed part of their prejudices and stereotypes to their texts, misjudging the tribal organization and their cultures. What is more important, the first ethnographers employed natives as a mirror to understand their own societies believing that sooner or later-for being strong or weak-their culture will wither way. This sentiment of ethnocentrism associated to a culturally-enrooted paternalism that symbolized the European ideals of enlightenment as the North to follow. While ethnographers should be "there" in order to collect much as possible from the hosting culture, colonial officers employed ethnographies to understand better what they dubbed as "noble savages." As Mary Louise Pratt puts it, the imperial eyes not only were decided to colonize the Other but also introduced the needs to do it. While the imperial center marked the different-alterity, it avoided being marked. From its onset, literature, and Science were of paramount importance in the ideological construction of the European (colonial) Matrix (Pratt, 2007). The colonial traveler sets the pace to the modern tourist-though echoing Pratt, the same cultural mandate persists. In tourism fields, for some reason which is very difficult to precise here, remained indifferent from the benefits of employing ethnography. Today's tourists and their obsessive quest for authenticity and heritage consumption emulate much of this colonial discourse. For that reason, being there-to understand-still plays a leading role in culture consumption. Unfortunately, tourism-related scholars have paid little attention to the connection of tourism and colonialism (Caton and Santos, 2008; Tzanelli, 2006, 2007, 2016). To fulfill this gap, the present chapter deals with ethnography, its definitions and historical evolution, as well as the challenges for the years to come.

Etymologically, the term ethnography stems from the Greek voices *ethnos (folk) and graphos (I write). One of the first ethnographers was Tacitus, who in his book De Germania, described with accuracy the habits and customs of ancient Germans. The General Julius Caesar documented his stay in Gaul, as well as the different challenges as long as his campaign. There reveals that there is an uncanny interplay between ethnography and imperialism. The exemplary travelers need to create an image of the (uncivilized) "Other" which is functional to their own idea of supremacy. Such a point evinces* that ethnography can be used as a document which interprets, and re-appropriates the "Other" according to the eyes of the fieldworker. In these terms, the notion of alterity seems to be fitted against selfhood as antithetical elements in the same game. The tensions between ethnographers and natives were widely discussed in the literature, but less has been said on how ethnography can be introduced in tourism studies. The present chapter focuses on the advantages and disadvantages of ethnography as a method which is recently introduced in tourism fields. One of the errors at the time of conducting ethnography associates to the fact that researchers misjudge-for some reason-the application of interviews-as qualitative techniques-with ethnography. The professional-ization of anthropology, which was accompanied by the growth of courses and publications, dates back to the 18th century. This encouraged many fieldworkers to launch towards the colonies geographically located in Australia, Africa, and Asia. With the army-forces and the trade and busi-ness relations, anthropologists were vested of a sacred authority which was based on scientific reason. Originally, ethnography was not separated from literature. To be more exact, the separation took place just after the Boasian anthropology struggled for a more standardized technique to dissect the cultural asymmetries. To some extent, ethnography, and colonialism were historically intertwined. This does not mean that anthropology or ethnology legitimized the conquest and arbitrary annexation of over-sea territories but also it contributed to the construction of alterity as a mirror to the European civilization. In the same way, ethnographers traveled long distances to meet the non-western "Other," nowadays tourists seem to perform a similar role. Valene Smith-in her landmark book *Host and Guests: the anthropology of tourism*-scrutinizes the nature of tourism and its profound changes for the community. As a subtype of leisure, tourism ritualized so that citizens escape temporarily from the daily pressures. As Smith adheres, the disposition to discover "Others' customs" -for different goals-equals tourism to the ideals of professional anthropology. Having said this, tourism anthropology has much to say respecting how this figure of "Otherness" is negotiated. What is more important, ethnography allows overcoming those obstacles or problems other methodologies failed to resolve. Of course, one might accept that over the recent years much controversy woke up revolving around ethnography. At a first look, as a qualitative method ethnography does not suffice to study large samples of population or sometimes has some problems for being replicated beyond the selected unit of observations. Key informants often see a part of reality and some valuable participants avoid taking part in the study. Secondly, the attitudes observed even in the same group can change in the threshold of time. In her book *Inalienable Possessions,* Annett Weiner (1992) obtains a radically different diagnosis-in her fieldworks in Australia-than Malinowski in his days. This suggests that-behind its immense efficacy to understand complex matters-ethnography still has its limitations. In the current chapter, we interrogate furtherly in the different applications of ethnography in tourism-related studies as well as sharing our own experience as fieldworkers in the past. In so doing, we review a short historical background from where ethnography departed. Secondly, we debate to what extent ethnography helped some well-reputed scholars to approach their object of study. As the backdrop, the French ethnologist Marc Augé (1994) calls the attention of a radical transformation of the city. Over the past centuries, anthropologists, and fieldworkers launched to over-seas territories to be in direct contact with the non-Western Other." These travels not only shed light on the lifestyle of natives as well as the formation of their hierarchies and institutions, but re-activated a tradition based on travel writing as a main literary genre. Nowadays, natives live in our cities and the notion of the alterity was radically transformed. The borders of the city have been enlarged while the fear of strangers was incorporated as mediation between lay-citizens. Last but not least, the challenges and problems of ethnography to be applied in tourism are mentioned in a final section.

4.2 A SHORT HISTORY OF ETHNOGRAPHY

There is an apparent disillusionment for the results of quantitative methods in the social sciences. Above all, naturalism, and positivism portrayed a world-in the strict sense of the word-which has been ruled by mechanistic laws. Based on the dialectics of rationality, positivism proposes the society as an immutable object which would be in some way studied objectively by an external researcher. Instead, ethnography confronts directly naturalism starting from the point that social reality is not enmeshed in a causal relationship but in micro-social interactions which are culturally interpreted from different angles. Such interpretations depend upon time and culture and are constantly shifting according to the negotiated values of the community (Hammersley and Atkinson, 2007). While positivist epistemologists emphasize on the needs of isolating all variables to be empirically tested at the time the sample is not contaminated by the influence of researcher, ethnography advocates by a process of reflexibility where the resulting experiences between observer and observed person are systematically fused into a hybrid text (Jack, 2007). In a nutshell, ethnography should be understood as one of the most important of the qualitative methods which combine different techniques to offer a systematic understanding of the observed culture. Its origins can be traced back to the inception of ethnology and anthropology (Agar, 1996; Clifford and Markus, 1986). The technique consists in observing and interacting with key-informants to decipher particular issues which not only are profoundly-ingrained in the core of the culture but also cannot be understood with other techniques. Neither trained in the anthropological knowledge nor in the art of field-working, the first ethnographers were lawyers moved by understanding "non-Western cultures" before their final disappearance. As Marvin Harris (2001) brilliantly observed, European paternalism envisaged the (non-western) native as an "Other" uneducated and uncivilized, often living outside the civilization or unfamiliar with the civilized customs. Nonetheless, the European intelligentsia strongly believed that this wild "Other" was not corrupted by the presence of civilization. In this respect, he lived freely and immune from the corruption of civilization and greed. Meanwhile, European colonizers developed an uncanny fascination for the native cultures, which alternated from the needs of expanding the European civilization to an unknown world, to the exercise of brutal violence to domesticate the aboriginals, who resist to the enlightened reasoning. Starting from the premise that the pre-modern Europe simply disappeared before the rise and evolution of industrialism, ethnographers strongly imagined the fate of Indians was sealed. Like Europe in the past, aboriginal cultures will be withered away before the industrial or capitalist logic. Marvin argues convincingly that ethnographers were encouraged to gather and collect cooking-utensils, shields, stories, lore, and all which will be destroyed by Europeans in their paths. Museums were the natural recipient of all these pieces delivered to the greater metropolis. Paradoxically, ethnographers' fieldwork notes were systematically read and used by the colonial governors to understand and anticipate the next aboriginal attack. Although ethnography and anthropology in particular were not responsible by the cruel crimes of the colonial rule, no less true seems to be that both were consistent with the colonial period, Marvin adds. Aside from this, ethnography proved to be today an efficient method to unravel the complex meeting between the selfhood and the alterity. What ethnography teaches seems to be linked to the fact of understanding our cultural values through the lens of the "Other," who looks different to us. By opposition, ethnography constructs-after living with native long period of time-a bridge-if not a window-to interpret the observed behavior within the cultural values where they operate. To put the same in bluntly, echoing Clifford Geertz (2008), winking may be very well interpreted as an invitation or simply it is an instinctive reflex. This means that actions and human behaviors cannot be mechanistically understood beyond the cultural ethos where they are shaped. As one of the fathers of the modern anthropology, Bronislaw Malinowski considered ethnography as an all-encompassing instrument that reflected empirically biological, social, and economic branches of the human constitution. The participant observation-accompanied by a long stay-provided some evidentiary bases where the ethnographer experienced-in his skin-the natives' lifestyle (Malinowski, 2002). Over the years, ethnography has varied to multiple versions as well as applications such as self or auto-ethnography (Reed-Danahay, 1997; Delamont, 2007), confessional ethnography or life-stories (Richardson, 2003), feminist ethnography (Stacey, 1988; Skeggs, 1994, 1988), critical ethnography (Anderson, 1988; Thomas, 1993), and recently virtual ethnography (Hine, 2000, 2008). At first look, the classic approaches frames ethnographers as a third person who obtains essential information by interviewing selected members of the group, who are dubbed as "key-informants." Based on certain objectivity, this insight develops an accurate diagnosis of the situation. In some cases, where informants cannot be revealed of the real goals of the research, it changes to a self-ethnography (Collins and Gallinat, 2010). It is noteworthy that ethnographers are not obliged to show what are their intentions and their role remains covered.

This happens in topics that confront with the legality of law and order such as criminality, suicide, or even terrorism. In these occasions, ethnographers can be pressed to reveal their key informants which contradict their own ethical code (Denzin, 2003). Rather, the critical ethnography advocates for the study of oppressed groups or communities which were historically pushed to live in a climate of inequality and domination. As Jim Thomas (1993) noted, critical ethnography should be defined as a sort of technique usually employed to explore cultures, actions as well as the political hierarchies of human groups. The main goal of critical fieldworkers consists in deciphering covert issues otherwise remain inexpugnable for lay-persons. In so doing, the idea of common-sense or status quo should be at best defied. At the bottom, while classic fieldworkers devote their efforts to understand what they observe, critical ethnographers, instead, work to change the given situation (using their work for emancipator goals, citing Thomas (p. 5).

Doubtless, one of the most compelling ethic issues seems to be related to the credibility ethnography has (first of all for other positivist epistemologists). What is important to discuss is to what extent the findings obtained by ethnographers can be replicated in other human groups or universes-as positivism objects- or even if the same group keep the same attitudes and habits in the threshold of time. As an art, ethnography was seriously questioned by the supporters of quantitative methodologies. In this vein, (Madison, 2011) argues that ethnography should move in the constellations of specific rules and ethics. As a story to be told to others, ethnography gives fresh and valid information on deep-seated matters which are covered by the ruling group (Madison, 2011). Legitimacy plays a crucial world, cementing the power of top-ranked members of the community over low-ranked others. The established ideology, which means the reason why some should be dominated by others, is the first obstacle the critical ethnographer daily face in his fieldwork. Madison goes on to say:

"What does it mean for the critical ethnographer to resist domestication? It means that she will use resources, skills, and privileges available to her to make accessible-to penetrate the borders and break through the confines in defense of-the voices and experiences of subjects whose stories are otherwise restrained and out of reach" *(Madison, 2011, p. 6).*

Having said this, what seems to be equally important is that ethnography should not be cognitively dissociated from the hierarchies of power or the economic cycles of production. Ethnography seems to be more than a source of scientific knowledge; instead, it is a source of politics. In normal conditions, ethnographers certainly captivate the attention of the tribal chief who needs to keep their hegemony in the group, or from the lowered-ranked members who manifest their discontent with the ruling group (Donham, 1999; Foley, 2002; Madison, 2011; Hart, 2004). In a seminal book, which entitles *Ethnocentrism: ethnography, history, and literature,* Arnold Krupat (1992) alerts that ethnography and anthropology base their observations and findings in the figure of the frontier. By dividing a constituent belonging to here and there, them, and us, the frontier leads gradually to *ethnocentrism.* Krupat acknowledges that European ethnologists not only re-appropriate the native's voice but also re-write their discourses filling their core from an ideological content. Although the process of reflexibility, which means the ability to put oneself in the other's place, is essential for a good ethnography, ethnographers commonly fall in binomial as West-East or other generalizations which obscure more than they clarify. In this respect, one of the paradoxes of ethnography consists in giving a voice to the native over-looking its essential features. The word natives and so to speak, aboriginals involve countless forms of organizations and cultures which are often singled out by anthropology to a one-sided interpretation. Seeing the World in dichotomized oppositions means legitimizing the idea that hapless Indians suffered the cruelty of European Lords, and vice-versa, that all discoverers were hatred-filled maniacs thirsty of Gold and Silver. Krupat straddles the idealized vision of Todorov and other cultural analysts, as well as the academic position of multiculturalists who supports the role of ethnicity over other cultural values. After all, Indians never identified by biological assets, rather they belonged to linguistic compatibilities. They never asked to what nation or race are you belonging? But what language do you speak?

As the author eloquently writes:

"At the risk of elaborating the obvious, I want to state clearly that my own critique of dichotomous logic as inadequate to the actual complexities of cultural encounter in history is not at all intended to endorse Todorov's or anyone else' notion that is possible somehow to proceed neutrally, avoiding not merely the reductionist practices

of scapegoating and victimizing but as well the moral political implications of any situate discourse" (pp. 20, 21).

The above-cited excerpt shows two important assumptions. On one hand, the notion of history stems from Western rationality which precedes it. The idea of chronological time was never certainly shared by aboriginal cultures. On another, Krupat coins the term *ethnocriticism* to shift the dialogical dynamic of Western reasoning while rejecting all forms of Manichean narratives. This raises a more than a pungent question: what would be the uses and advantages of ethnography in tourism-related research?

For some reason, which is very hard to precise here, tourism-research-from its inception-evolved towards an economic-centered paradigm, which means a new family of theories based on the economic effects of tourism at the destination. Although some managerial disciplines adopted ethnography as a valid source of investigation, this material paradigm prioritized quantitative over qualitative methods. As stated, researchers focused on the administration of open or close-ended questionnaires to understand the complexity of the industry. Any tourist as well as its expectances and preferences-probably excluding other actors-situated as the main object of study of the economic-based theory. Over recent decades, some critical voices have alerted on the risks of approaching what tourists say as the only way for accessing the fieldwork (Tribe, 2010; Rakic and Chambers, 2011). In other cases, the paradigm failed to give an all-encompassing model of what tourism is. Most certainly, the questionnaires or informal interviews are applied on hub, airport, and bus stations without a chronological follow-up of what the interviewee feels years later. Given this, the emotionality and the theory of sensibilities are systematically ignored in the process of knowledge production. As signpost towards new answers, we add that ethnography showed how sometimes interviewees lie or distort their responses while in others-at the best-they simply are unfamiliar with their inner-most emotional world. For example, what would reply a gangster or a prostitute whether we ask: what is your profession?

One might speculate that the former will answer a businessman while the latter would reply: secretary. Under some conditions, the interviewed person is interrogated by its prejudices, believes, and expectances which are culturally molded by the status quo. What can be said or not depends upon complex social and inter-individual forces operating in a complex scenario. Aside from the discrepancies and critical points revolving around ethnography, its benefits orient to give an evidentiary basis on the above-mentioned methodological limitations of positivism and the economic-centered theory.

4.3 ETHNOGRAPHY IN TOURISM

Since ethnography has been historically the tug of war of anthropology, its arrival was accompanied with the anthropology of tourism. Insofar as ethnographers traveled long vast distances to meet with the alterity, today tourists launch to colonize other cultures and territories through their gaze (Bruner, 1989; Urry, 2002; Urry and Larsen, 2011). In consonance with Nelson Graburn (2002), it is safe to speak of an "ethnographic tourist" as an emerging technique aimed at overcoming the obstacles ethnography often meets in urban contexts. Per his viewpoint, ethnographers in the city have many problems to classify and select vital information without mentioning the change of attitudes and habits. The ethnography of tourist, as Graburn puts it, the cultural similarities between fieldworkers and city-dwellers create some difficulties in the ethnographic task. As "Other," tourists epitomize the role of natives in classic anthropology. One of the pioneering and seminal (auto) ethnography in tourism fields was conducted by Dean MacCannell (1976) in his book *The Tourist: a new theory of leisure class.* Centered on continuing the debate left by T. Veblen on leisure class, he toys with the belief that tourism has occupied the same role of Totem in the tribal society. He conducts an auto-ethnography in Walt Disney World Epcot Center providing with fresh observations elaborated during his sojourn. It is important not to lose the sight of the fact that MacCannell attempts to avoid formal interviews with his key-informants. Rather, he prefers to see and hear keeping a passive role. We are educated to believe that the "Other" has much to say, and sometimes, a good ethnography implies only hearing.

As the previous argument is given, other interesting approaches are the book *Etnografia bajo un espacio turistico (ethnography on a tourist place),* a text authored by Spanish anthropologist Antonio Nogues Pedregal (2015); Phillip Vannini's text *Ferry Tales* (2015); and *Tourism Imaginaries* (Salazar and Graburn, 2014). From different angles and moved by different aims, each book concentrates efforts in stripping away politics the veil of sainthood. At a closer look, the tourist space is a place of politics, interaction, rivalry, and conflict. Nogs Pedregal centers his case in Zahara de los Atunes city (Spain). The adoption and evolution of tourism as a main industry not only accelerates some irreversible changes but also re-structures the kinship relations, as well as the disputes for power and the accessibility to the new resources that invariably this industry generates. To some extent, tourism reinforces the previous material asymmetries-among clans and groups-laying the foundations to a new financial dependency. Phillip Vannini in his book *Ferry Tales: Mobility, place, and time on Canada's west coast*. The role of Vannini is pretty different than his Spanish colleague. Vannini focuses on what exerts dubs as *mobile ethnography* during a trip in the ferry. His thesis is that accidents alter our conceptions around normalcy, as well as security. The accident interrogates furtherly on the definition of safety for social imaginary. The supremacy of technology changes not only the chronological time but also leads to seeing failures in the function of machines as improbable. Since our biographies are determined by death, culture organizes its essential structure around the figure of the accident. Vannini contends that mobility is certainly established and controlled by capital and its rapidity of expansion. While politics is aimed at encouraging the rights to access mobilities in some privileged actors, it immobilizes others who lack voice and representation. To solve these short-circuits researchers should conduct innovative ethnography collating silenced stances. In doing so, our ethnographer recurs to the "more than-representational theory," which focuses on the significance in practices and senses in sharp opposition to symbols and cultural codes. This new methodology re-questions the existent definition of technology (mainly referred to as machines and instrumentality).

Last but not least, Noel Salazar and Nelson Graburn (2014) edit a book under the title *Tourism Imaginaries*. Although each chapter holds different positions, there is a common-thread argument to follow. Editors evince a strong intention not to approach tourists as the only actor of a global system, introducing ethnography to interview and understand other actors. The conflict-not consumption-is the main object of study of all chapters that form this book. Tourist imaginaries are drawn to mediate between culture and natives. Imaginaries are conceptualized as "transmitted representational assemblages interacting with people and landscapes towards a world-shaping cosmology. Ethnographers should penetrate these imaginaries to tackle off not only the ideology behind politics but also the multiple shapes and paths it takes in the community. The tourist imaginary remains hidden to the tourist gaze, so ethnographers should grasp the operating logic and its possible manifestations observed as long as the fieldwork. Sometimes there is certain dissociation between reality and what people finally imagine. Let's remind readers that in fieldwork we conducted in Cromañon Sanctuary, a dark tourism spaces where 194 persons lost their lives-in a nightclub fire occurred on 30 December of, 2004, a female 24 years old met us to be interviewed. After long hours and days of recording and transcript, we realized, unfortunately, a great part of the information was false, imagined or distorted by the interviewee. The content of the interview was rich in content as well as expressed the common discourse available the Argentinean social imaginary-just after the tragedy. This young man has something important to tell us while framed his discourse in the social discontent of the citizenry against the politicians. Of course, the interview was not discarded- and even though it was not published-it served us to reach more valuable biographies and life-stories. This reveals that sometimes what people say is not the truth but it articulates a much dense plot which sheds light on the shadows of secrecy. Identity, at least anthropologically speaking, fragments in two sides. The image of the id others has and the image of the id I normally imagine.

4.4   CHALLENGES AND DILEMMAS OF ETHNOGRAPHY IN TOURISM FIELDS

Ethnography, as discussed in the present chapter, takes many forms and shapes. In its origin, it was dated back to the rise and expansion of colonial powers (as it was described in the earlier sections). Based by colonial authorities to understand their enemies, ethnography was inevitably associated to the colonial cruelty. This does not mean that ethnographers were conducive to support colonialism. They adventured to travel to overseas territories to understand the *Non-western Other* in sharp contrast to the civilized western reasoning. In this vein, no fewer problems emerged at the time of conducting fieldwork. Instead to what researchers believe, ethnography is something more complex than simple formal or informal interviewees organized to test an underlying hypothesis. Although, any fieldworker starts from a hypothetical scenario which was cemented through the reading of previous chapters, studies, books or works in the matter. Once the problem is correctly formulated, the fieldworker should select the unit of analysis as well as its role in the ethnography. As we have discussed, an overt role indicates that the observed group is aware of the research goals while a covert role, most probably in cases of criminality, racism, or terrorism, needs key-informants to keep certain ignorance respecting the orientation of the research. This moot-point opens the doors towards a new discussion in the fields of ethics. To what extent is ethical to study human beings without their consent? Or what is worse, if I reveal overtly my aims to a white supremacist in what way he will manifest his deep feelings?

An answer to these questions is not an easy task but it depends on many factors. The theme as well as the research design not only punctuates on the role of the ethnographer but also the number of visits and stay time in the group. Classic ethnographers resided long period, but this is not rentable any-longer. Today, ethnographers live with natives only weeks or months. Still further, Western institutions which often fund fieldworkers' stay may exert influence in the topic and how it should be framed. For many natives, ethnographers are representant of the nation-state and for that considered unworthy. Third, the role of empathy occupies a central position for the fieldwork to arrive at a safer port. To put the same in bluntly, the neces-sary trust to create a bridge between the fieldworker and key-informants request several efforts and time. Not all members of the group are willing to take part in the interview, as we have mentioned in the earlier sections. Once this stage is successfully achieved, the ethnographer should start his work. In so doing, preliminary informal interviews take place. This helps the researcher to explore in an initial stage to organize the variables for designing more robust formal interviews. The tape recording cannot be used in all situations. For example, themes associated with racism or criminality cannot be performed with a recorder on the hand. On those questions where there is a strong moral repudiation, key-informants adapt their replies to avoid legal issues. When a considerable number of formal interviews are reached, the ethnographer can select three or four story-lives to support his main theses. In practical terms, though there are many methodologies to approach key informants, snowball technique is the more recommendable. This consists of asking each interviewed to recommend other potential participants to recruit. At the time the fieldworker returns to home, a final report should be written, in a clear and polished style. Some methodologists caution that this represents a serious risk for the objectivity of the report since the native's voice can be misjudged, ill-interpreted or involuntarily distorted. At the time the ethnography ends as soon as the final report is written the better. Last but not least, a crucial dilemma is given when ethnographer should publish the results. Because of ethical concerns, the real name of key-informant should be altered or changed but, in some extent, they exert a radical pressure on what the ethnographer will write. Today, the process of globalization-adjoined to the digital technologies-facilitates natives and key-informants to access the research in hours. This process of flexibility seems to be a great challenge posed on the ethnography. When the topic associates to illegal activity or terrorism, to set an example, officials, and governments press ethnographer to reveal their key informant's identity. Ethnographers may be very well imprisoned whether he or she decides not to share the requested information. Not only the moral condemnation on some topics make ethnographer's job very difficult but also opens a gap which is filled by pseudo-experts. As Luke Howie brilliantly stressed, fear, anxiety, and obsession for terrorism lead Western spectatorship to accept ideas and policies otherwise would be rejected. Not only governments press ethnographers to give voluntarily their findings, but also terrorism is esteemed as an extreme and serious crime. Hence, the media and TV programs invite some voices who-far from conducting serious research-support the dominant discourse elaborated by the ruling elite (Howie, 2012).

4.5 CONCLUSION

The history of anthropology reveals two significant things. On one hand, the needs of discovering while protecting the "non-Western Other" before its final disappearance. The European order not only expanded to colonize the world but also portrayed natives as naïve children, uncultivated and uneducated who deserved to be civilized. This civilization consisted in introducing natives into the ideals of western rationality and the Enlightenment. In the process, anthropology played a leading role providing colonial officers vital information to domesticate (control) the locals. As Mary Louise Pratt puts it, the imperial eyes not only were decided to colonize the Other but also created the needs to do it. While the imperial center marked the different-alterity, it avoided to be marked. From its onset, literature, and Science were of paramount importance in the ideological construction of the European (colonial) Matrix (Pratt, 2007). The colonial traveler sets the pace to the modern tourist-though echoing Pratt the same cultural mandate persists. In tourism fields, for some reason which is very difficult to precise here, remained indifferent from the benefits of employing ethnography. To fulfill this gap, the present chapter dealt with ethnography, its definitions and historical evolution, as well as the challenges for the years to come.

KEYWORDS

- **anthropology**
- **ethnocentrism**
- **ethnographers**
- **ethnography**
- **tourism and colonialism**
- **tourism imaginaries**

REFERENCES

Agar, M. H., (1980). *The Professional Stranger: An Informal Introduction to Ethnography* (p. 117). Academic press, New York.

Anderson, G. L., (1989). Critical ethnography in education: Origins, current status, and new directions. *Review of Educational Research, 59*(3), 249–270.

Augé, M., (1994). *Hacia una antropología de los mundos contemporáneos (Towards an Anthropology of contemporary Worlds)*. Editorial Gedisa, Barcelona.

Brewer, J., (2000). *Ethnography*. McGraw-Hill Education, London.

Bruner, E. M., (1989). Of cannibals, tourists, and ethnographers. *Cultural Anthropology, 4*(4), 438–445.

Caton, K., & Santos, C. A., (2008). Closing the hermeneutic circle? Photographic encounters with the other. *Annals of Tourism Research, 35*(1), 7–26.

Clifford, J., & Marcus, G. E., (1986). *Writing Culture: The Poetics and Politics of Ethnography*. University of California Press, Berkeley, CA.

Clifford, J., (1988). *The Predicament of Culture: Twentieth-Century Ethnography, Literature, and Art*. Harvard University Press, Cambridge, MA.

Collins, P., & Gallinat, A., (2010). *The Ethnographic Self as Resource: Writing Memory and Experience into Ethnography*. Berghahn Books, London.

Delamont, S., (2007). Arguments against auto-ethnography. In: *Paper Presented at the British Educational Research Association Annual Conference* (Vol. 5, p. 8).

Denzin, N. K., (2003). Performing [auto] ethnography politically. *The Review of Education, Pedagogy & Cultural Studies, 25*(3), 257–278.

Dixon-Woods, M., (2003). What can ethnography do for quality and safety in health care? *BMJ Quality & Safety, 12*(5), 326–327.

Donham, D. L., (1999). *Marxist Modern: An Ethnographic History of the Ethiopian Revolution.* University of California Press, Berkeley, CA.

Foley, D. E., (2002). Critical ethnography: The reflexive turn. *International Journal of Qualitative Studies in Education, 15*(4), 469–490.

Geertz, C., (2008). Thick description: Toward an interpretive theory of culture. In: *The Cultural Geography Reader* (pp. 41–51). Routledge, London.

Gobo, G., (2008). *Doing Ethnography.* London, Sage.

Graburn, N., (2002). The ethnographic tourist. In: Dann, G., (ed.), *The Tourist as a Metaphor of the Social World* (pp. 19–39). CABI, Wallingford.

Hammersley, M., & Atkinson, P., (2007). *Ethnography: Principles in Practice.* Routledge, Abingdon.

Hammersley, M., (2018). What is ethnography? Can it survive? Should it? *Ethnography and Education, 13*(1), 1–17.

Harris, M., (2001). *The Rise of Anthropological Theory: A History of Theories of Culture.* AltaMira Press, Chesnut Creek AR.

Hart, G., (2004). Geography and development: Critical ethnographies. *Progress in Human Geography, 28*(1), 91–100.

Hine, C., (2000). *Virtual Ethnography.* Sage, London.

Hine, C., (2008). Virtual ethnography: Modes, varieties, affordances. In: Fielding, N., Lee, R., & Blank, G., (eds.), *The SAGE Handbook of Online Research Methods* (pp. 257–270). London, Sage, London.

Howie, L., (2012). "Witnessing terrorism". In: *Witnesses to Terror* (pp. 155–175). Palgrave Macmillan, London.

Hymes, D., (2003). *Ethnography, Linguistics, Narrative Inequality: Toward an Understanding of Voice.* Taylor & Francis., London.

Jack, G., (2007). International management and ethnography: What and why? *Ethnography, 8*(3), 361–372.

Krupat, A., (1992). *Ethnocriticism: Ethnography, History, Literature.* University of California Press, Berkeley CA.

MacCannell, D., (1976). *The Tourist: A New Theory of the Leisure Class.* University of California Press, Berkeley, CA.

Madison, D. S., (2011). *Critical Ethnography: Method, Ethics, and Performance.* Sage publications, London.

Malinowski, B., (2002). *A Scientific Theory of Culture and Other Essays* (Vol. 9). Psychology Press, London.

Nogues, P. A., (2015). *Etnografía Bajo un Espacio Turistico.* [Ethnography of a tourist space]. Universidad de la Laguna Press, El Sauze.

Pratt, M. L., (2007). *Imperial Eyes: Travel Writing and Transculturation.* Routledge, Abingdon.

Rakić, T., & Chambers, D., (2011). *An Introduction to Visual Research Methods in Tourism* (Vol. 9). Abingdon, Routledge.

Reed-Danahay, D., (1997). *Autoethnography*. Berg, New York, NY.

Richardson, L., (2003). Writing: A method of inquiry. *Turning Points in Qualitative Research: Tying Knots in a Handkerchief,* 379–396.

Salazar, N. B., & Graburn, N. H., (2014). *Tourism Imaginaries: Anthropological Approaches*. Berghahn books, Oxford.

Skeggs, B., (1994). Situating the production of feminist ethnography. In: Maynard, M., & Purvis, J., (eds.), *Researching Women's Lives from a Feminist Perspective* (pp. 72–92). Taylor & Francis, London.

Smith, V. L., (2012). *Hosts and Guests: The Anthropology of Tourism*. University of Pennsylvania Press, Philadelphia, PE.

Stacey, J., (1988). Can there be a feminist ethnography? In: *Women's Studies International Forum*. Pergamon, Oxford.

Thomas, J., (1993). *Doing Critical Ethnography* (Vol. 26). Sage, London.

Tribe, J., (2010). Tribes, territories and networks in the tourism academy. *Annals of Tourism Research, 37*(1), 7–33.

Tzanelli, R., (2006). Reel western fantasies: Portrait of a tourist imagination in the Beach 2000. *Mobilities, 1*(1), 121–142.

Tzanelli, R., (2007). *The Cinematic Tourist: Explorations in Globalization, Culture and Resistance*. Routledge, Abingdon.

Tzanelli, R., (2016). *Thana Tourism and Cinematic Representations of Risk: Screening the End of Tourism*. Routledge, Abingdon.

Urry, J., & Larsen, J., (2011). *The Tourist Gaze 3.0*. Sage, London.

Urry, J., (2002). *The Tourist Gaze*. Sage, London.

Vannini, P., (2012). *Ferry Tales: Mobility, Place, and Time on Canada's West Coast*. Routledge, Abingdon.

Weiner, A. B., (1992). *Inalienable Possessions: The Paradox of Keeping-While Giving*. University of California Press, Berkeley, CA.

CHAPTER 5

Tourism, Conflict, and Conflictivity: Is Tourism Part of the Problem or the Solution?

ABSTRACT

German philosopher Immanuel Kant, in his prolific career, applauded the role of hospitality as an articulator of understanding and eternal peace among nations. To what extent he is wrong or not, following Hobbes and his realism seems not to be the central topic of this chapter. Rather, the chapter focused on challenges the tourism industry will come through in the next decades to achieve a durable peace. While the first section reviews a full-fledged part of the specialized literature, enumerating the methodological problems of economic-based paradigm, the second provides a snapshot of the promises and ethical problems of post-conflict destinations to regulate conflict. Finally, an agenda for the next years is debated. We start from the premise the evidence that supports tourism promotes peace is not consistent, but here further investigation is needed. There are interesting case-studies that show how tourism-given certain conditions-may very well boost local economies, disarticulating old rivalries, and leading the community towards a prosperous peace.

5.1 INTRODUCTION

The problem of conflict and violence is tough to be precisely captured even in social sciences (Kew and John, 2008; Nillson, 2012; Centeno, 2002). To what extent are humans naturally violent or prone to war? And what is worse, should conflict and violence lead to resolving the economic crisis? Of course, tourism seems not to be an exception. Over the years, scholars have systematically interrogated the position of tourism as the safeguard of peace and prosperity (Leslie, 1996; Pernecky, 2010; Becken and Carmignani, 2016). In their landmark book, *Ethnicity Inc.,* anthropologists Jean and John Comaroff alert that under some conditions, tourism triggers a long-dormant state of conflict among inter-ethnic groups. With a focus in the role of tourism in Africa and the US, they hold that some aboriginal groups which were historically relegated from the economic growth have recently adopted tourism as a valid mechanism to alleviate poverty. Their success implied new higher taxes by the side of government, paving the way for the rise of bloody ethnic cleansing and war (Comaroff and Comaroff, 2012). This text leads us to think that tourism and conflict have an ambiguous relationship that remains unexplored even to date (Salazar, 2006; Var et al., 1989; Weaver, 2011). In some cases, war or its negative effects may be a criterion of attraction for many tourists (Isaac, 2010; Wise, 2011; Suntikul, 2019). To a closer look, the industry of tourism was historically identified as peaceful because it needed political stability to survive. Having said this, tourism not only revitalized local economies but needs from peace to be expanded (Crouch and Ritchie, 1999). As sensitive to bad advertising and disrupting events, tourism certainly preserves the tourist security avoiding conflict and war (George, 2003; Hall, Timothy, and Duval, 2004; Cheng and Noriega, 2004). Experts agree that a stable condition for peace is the best course of marketing for a tourist destination (Scott, 2012; Beirman, 2002; Yoshida, Bui, and Lee, 2016). To put the same in bluntly, nobody wants to travel to a dangerous place where his or her life is in danger or runs some risk. Lea (2006); and Richards and Hall (2003) remind how the tourism industry lays the foundations towards a competitive atmosphere, which sooner or later develops peace and democracy worldwide.

But far from being acute in the observations, some scholars adopt a contrary position. As de Kadt (1979) eloquently showed, some nations which were subjugated by the colonial powers or culturally fraught in internal civil wars are less prone to develop successful tourism than others which never faced these post-conflict experiences. Given this controversy, the present chapter explores the intersection of tourism and conflict while adding some clarifications as a post-conflict destination as a murky concept. This chapter interrogates furtherly on the intersection of conflict and tourism while laying the foundations to understand the methodological problems and philosophical quandaries of post-conflict research.

The first section explores the literature that explains the evolution of tourism as a peace-builder. From its inception, tourism research focused on the opportunities in local communities to foster a durable peace through the articulation of tourist campaigns. Nonetheless, some voices emerged just after 2010 that have criticized the efficiency of tourism research for regulating war, terrorism, and war. These critical scholars led to re-consider the idea of post-disaster consumption as an efficient mechanism to boost affected local economies. This seems to be the main topic of the third section. Finally, we give some guidelines in a final section which entitles *the Agenda for the next decades.* Doubtless, scholars have not reached an agreement revolving around the role of tourism as a promoter of peace, but what is more important, there are ethical dilemmas in the field which should be re-visited.

5.2   UNDERSTANDING TOURISM AND CONFLICT

The assumption left in the introductory section of this chapter, which means that tourism promotes conflict, is far from being empirically validated but what it is important not to lose the sight of the fact that more than four decades elapsed after de Kadt's publication, but his seminal book holds fair and actual even to date. There is a dearth of new publications that focus on tourism as a vehicle towards peace and institutional stability (Litvin, 1998; Var and Ap, 1998; D'Amore, 1998; Cho, 2007). All these authors concentrate efforts and time in debating to what extent tourism is a vital force for peace, or simply it needs from stable conditions to flourish and survive (Leslie, 1996). As S. W Litvin puts it, tourism is often praised as a beneficiary of peace even if under some conditions, it succeeds in developing stable institutions, above all when a process of fair distribution of wealth is achieved. These voices are in agreement that tourism not only encourages peace but also needs it not to die. In consonance with this, Moufakkir and Kelly (2010) acknowledge that peace should be redefined adjoined to the concept of security. Each community elaborates its own tolerance to the uncertainness as well as the conception of what is a secure place. One of the limitations of current research associates to the understanding that conflict means a lack of peace. For this reason, the obtained results remain obscure and are not conclusive at all. In consonance with the earlier chapters, we argue that the economic-based paradigm exaggerated the role of tourism as a mechanism of peace and political stability, paving the way for the rise of literature that misjudged-if not condemned-conflict as the main threat to tourism industry. For Barlow (1988), or Rovelstad (1988)-one of the pioneering scholars in this field-tourism appears to open the doors to international coordination, which is the precondition towards a durable peace among nations. This contradicts Var and Ap, who argue that there is little evidence that suggests tourism regulates violence in the different levels of society. Centered on an applied-research, they conclude that there is an ecological fallacy in the studies of the field where tourism profes-sionals who are often interviewed by fieldworkers assume that tourism encourages long-lasting stability and peace. Methodologically speaking, as Var and Ap agree, there is needed further investigation-probably incorporating statistical information-to determine the impact of tourism in society. Echoing this point, Stephen Pratt and Anyu Liu (2015) accept tourists, who are in contact with other cultures, are psychologically open to new experiences which lead them to cement new bondage with locals. Following the GPI (global peace index) as their main source, they agree tourism is important for peace but not the determinant factor. Whatever the case may be, some mid-income destinations are more prone to peace and political stability than higher purchasing power destinations. The sample was originally formed to measure 22 qualitative and quantitative variables where perceived security and safety, as well as the degree of militarization in the destination, occupied a central position. Here two methodological problems arise. On the one hand, the concept of violence-conflict cannot be dissociated to other terms as prejudice, racism, or even tourist-phobia (a point to deal in the next chapter. On another, racism is daily maintained in secrecy according to the subordination of guest-host roles.

In earlier research, we have found robust evidence that the subordina-tion of hosts respecting guests play a leading role in the configuration of a covert-racism or hostility in the tourism industry. To set an illustra-tive example, over more than two decades (since the Beagle conflict), Argentinean tourism workers developed a covert hostility against Chilean tourists who visited massively the country since 2001. To some extent, history exerts considerable influence in the formation of hostile reaction to foreign visitors (Korstanje, 2011; Korstanje and George, 2012). The panacea of tourism as peace-builder was widely criticized by Sasha Pack (2006), who in her seminal book *Tourism and Dictatorship*, explores the connection of Franco's dictatorship with tourism. Moved to stimulate domestic consumption just after a civil war that decimated Spain, Franco turned his attention to the benefits of tourism. Although never democratized in Franco's regime, Spain obtained economic benefits from tourism not only consolidating his power but also notably growing. A similar aspect is formulated by Noel Salazar in his different works. The scientific evidence that confirms tourism encourages democracy and peace seems to be resting in shaky foundations. In this direction, he did an important step that war not necessarily exhibits an obstacle to the expansion of the tourism industry. Its resiliency proved to be an important stimulus to domestic consumption, and it is often adopted for dictatorships to placate the internal dissidence. He even cites:

> "On the other hand, certain acts of terrorism are specifically targeted at tourism. In conflict zones, tourists can be targeted because they are viewed as ambassadors for their countries, as soft targets, and often because of their "symbolic value as indirect representatives of hostile or unsympathetic governments." In countries like Burma, tourism brings international recognition and fosters an illusion of peace while providing foreign exchange to pay for arms which strengthen the military junta. In cases such as this, tourism fortifies the undemocratic regime whose members may benefit personally and politically from any increase in arrivals. Although controversial, a tourism boycott is believed to help diminish the access to such rewards and erode the foundations of the government, advancing the necessary political changes to establish peace"   (Salazar, 2006, pp. 325, 326).

In a nutshell, Salazar overtly says that low-politics fail to produce durable stability even adopting tourism as a main economic activity simply because, in its nature, tourism foments higher levels of competence and rivalry. In those cultures where there are old disputes the probabilities, the conflict surfaces are higher. In this point, Salazar equates to de Kadt's argumentation. History offers a fertile methodological ground to understand the relation of tourism and history (Salazar, 2006, 2010). In his book *Envisioning Eden: mobilizing imaginaries in tourism and beyond,* he starts from the premise that mobility theory has some problems to explain how the Otherness is represented in the Western imaginaries. Based on an ethnocentric discourse, Europeans applaud mobilities as an idealized value to follow, to mark the so-called superiority of some cultures over others. It is very hard to imagine tourism without violence because it activates when a cultural border is crossed or altered. To put the same in other terms, tourism, and mobilities break geographical borders, i.e., between the global North and South-while conflict articulated a counter-balance to the imagined border to be re-established. As Timothy (2019) puts it, it is common to think in tourism as a peace-builder, but far from this paradigm, scholars need to rethink tourism as a root of disharmony that under some conditions accelerates conflict and wars. This raises some pungent questions such as, are conflicts inevitably entwined to tourism? Is tourism a resilient industry?

5.3   WHAT WE KNOW ABOUT POST-CONFLICT DESTINATIONS?

Doubtless, some voices have alerted that the turn of the century accelerated some external and global risks, such as terrorism, political violence as well as disasters or virus outbreaks-like COVID-19-which is gradually harming the tourist system. This era is characterized by the urgency to protect the tourist system in order for the global governance which means the capacity of political powers to keep stability, does not been seriously affected (Mansfeld and Pizam, 2006; Korstanje and Tarlow, 2012; Tarlow, 2014; McKercher and Chon, 2004). The notions of security, risk perception or safety played a leading role in the fields of tourism management. The precautionary principle was employed to anticipate and potentially eradicated those risks that hurt the organic image of the tourist destination. The same applied for conflict (Laws, 2005; Timothy, 2006). The theory of risk perception, which was borrowed from psychology, was engulfed as the tug of war of tourism researchers. The main goals of researchers were to locate those psychological or physical dangers that intervene in the decline of a certain destination. Based on a preemptive logic, these studies were enthused to adopt quantitative methods to prevent disasters, economic crises, or even regulating the potential internal conflicts (Blake and Sinclair, 2003; Ritchie, 2004; Glaesser, 2006). Having said this, the *crisis management* theory gradually becomes a paradigmatic model to be emulated by countless governments, policy-makers, and experts in the industry. It is important not to lose the sight of the fact that earlier than 2012, this academic wave was well-read and reputable worldwide. But here something was wrong. The rise of new more radicalized groups, like

Boko Haram and ISIS, accompanied with the acceleration of the green-house effects, showed not only the incapacity of tourism management to procure more security destination but also how the precautionary doctrine failed to give an all-encompassing model to predict future disasters. In fact, like Swine flu, SARS, or Ebola, today the COVID-19 is again placing the tourism industry between the wall and the deep blue sea. What the actions of terrorist cells revealed was the impossibility of precautionary doctrine to manage external or internal risks while posing serious challenges for the tourist system in the years to come (Seraphin, Butcher, and Korstanje, 2017). Over the recent years, some scholars not only exerted a radical criticism over the precautionary doctrine as well as risk perception to ensure safer destinations but also called the attention to the urgency to adopt new strategies to mitigate-not eradicate-the negative effects of conflict and violence in the society. For them, more resilient destinations should be adapted to the new times, and in so doing, experts need to expand the current understanding of conflict and how they affect the system. The main thesis of these critical approaches rested on the fact what tourists experienced, or how the tourist experience is formed do not tackle the problem, only speak of how tourists perceive some destination (Walters and Mair, 2012; Chew and Jahari, 2014; Seraphin, Korstanje, and Gowreesunkar, 2019). Not surprisingly, all these studies emerged as a counter-response to the open questions left by the risk perception paradigm just after 2012. Post-disaster and post-conflict destination theories went in the opposite direction. This conceptual corpus is mainly moved by the premise risks cannot be avoided, but they can be adapted for the tourism demand finds new resilient destinations. The *post-disaster and post-conflict destination theories*, in this way, emphasize the urgency to enhance the resiliency of tourist destinations (Seraphin, Korstanje, and Gowreesunkar, 2019). In a prefatory introduction to the book *Tourism and Hospitality in conflict-ridden Destination,* Isaac, Cakmak, and Butler (2019) recognize that the difficult intersections between conflict and tourism lead scholars to interrogate to the dominant discourses in tourism fields mainly tended to ponder security and safety as unquestionable mandates.

"Studying the complex intersections prevents the hegemonic discourses of global security for completing itself, stabilizing its boundaries and securing a totalized presence. These hegemonic discourses create a separation between socio-political conflicts

and tourism through recurrent reminders of the necessity of safety
and security in tourism." (Isaac, Cakmak, Butler, 2019, p. 1).

Echoing the above-cited excerpt, an increasing number of studies have focused on the post-conflict destination management as a potential alternative to the different disputes that should be regulated for the correct well-functioning of the industry. In this vein, post-disaster or post-conflict destinations rest on allegories or memorial narratives which are constituted once the event happened. As Naef Patrick (2019) eloquently observes, tourism not only helps in the construction of a shared (negotiated) memory but recreates the conditions for a bridge that leads towards reconciliation. With focus on the case study of Colombian Narcos, she holds the thesis that memorial entrepreneurs play a leading role-as individuals-looking for the complete understanding of the historical (bloody) events which shocked the society in the past. This dissonance between past and present is known as dissonance heritage (Hartmann, 2014; Roberts, 2018; Biran and Buda, 2018). With this in mind, Anne Marie Van Broeck (2019) analyzes the case of Moravia, Medellin (Colombia). Through the figure of *Phoenix tourism, a neologism recently coined by some experts*, which corresponds with a new segment of tourists interested to visit post-conflict sites. She finds that once the Colombian government agreed on a vexed post-agreement with FARC (revolutionary armed forces of Colombia) in 2016, tourism upsurged as a potential economic factor to boost the local economy. To some extent tourism made from the city a more attractive destination accelerating important demographic shifts and urban transformation, but at the same time, some conflictive situations surfaced. Locals were originally discontent with external tour operators who packaged and sold their externally-fabricated narratives about terrorism in Colombia. They feel not only they were relegated from the economic benefits of tourism, but also, they were unilaterally to tell a story they never lived. This point a more than interesting question related to the ideological power of dark consumption or post-conflict discourses?

Similar remarks can be found in other works such as Thanatourism and cinematic representations of risk which is authored by Rodanthi Tzanelli, and Sather-Wagstaff's book Heritage that Hurts, two well-read editorial projects that punctuate the negative core of post-conflict destinations. Tzanelli and Sather-Wagstaff share a similar position respecting the accuracy of heritage entrepreneurs or policy-makers to tell the true story to visitors. Of course, the arriving tourists there wish to hear the story behind the sad event, but such a story may be very well fabricated protecting some interests. This means not only the dominant narrative distorts the previous background behind the event, but also the process of rememorizing does not explain the causalities that led the community to the state of war or conflict. Joy Sather-Wagstaff (2018) acknowledges that tragic events wake up inherent human solidarity for the victims putting people in equal conditions before death. This sentiment of reciprocity straddles borders, nationalities, religions, and cultures. In this respect, she is focusing on the Ground-zero and 9/11 as a foundational event for the US. In the sad days after 9/11, the US received global support even from distant or culturally-different nations. This situation rapidly changed when Bush's administration monopolized the political discourse launching to an ally to head two invasions to the Middle East. As she notes, while the empathy with the Other is a natural condition that all humans share, the political manipulation of sadness corresponds with the rise of she dubs as a *dark heritage (or in his terms heritage that hurts)*. In fact, she goes further, heritage exhibits a political manipulation of a historical event, an event never is told or rememorized as it occurred. As introduced, Tzanelli (2018) has some points of convergence with Sather-Wagstaff but she remains in a different academic position. She is primarily motivated to respond to two interesting questions: why Western civilization has commoditized risk as a form of entertainment? and is West a sadist civilization?

As the previous backdrop, Tzanelli introduces her readers in a ground-breaking thesis, which suggests capitalism has mutated to a new stage where artists pivot in the drawing of a dark landscape that emulates the apocalypses. While these dark stories terrorize us, the world is far from being destroyed. Here is the first point of disconnection between Tzanelli and her colleague, Sather Wagstaff. For the later, dark tourism enacts positive solidarity forged in the tragedy, for the former it is commoditized as a form of the cultural entertainment industry. Steeping back to my previous formulated inquiry, Tzanelli sets forwards an explanation on the reasons why we are more prone to consume macabre sites. As she accepts, the allegories of death revolving around the cinema industry serve to protect and replicate the authority emanated from the ruling class. Tourism is far from disappearing, rather it is reinforced by a dark allegory where risks are inflated, commoditized, and disseminated for citizens to embrace the leading values of the nation-state. Inscribed in what anthropologists known as *a gift-exchange process,* dark tourism revitalizes long-dormant (colonial) discourses oriented to present the Non-Western Other as dangerous (probably epitomized in the figure of a terrorist, a monster, or a zombie). The film-related industries, in this vein, draw the risk to impose allegories that leads the audience to believe in capitalism as the best of feasible worlds. To cut the long story short, Tzanelli offers a fertile ground to believe that dark tourism deals with a representational past which are fictionalized and recreated to subordinate the Non-western and peripheral alterity. Dark tourism, which is fascinating the European tourists, says little on the historical responsibilities of European powers during the colonial rule (Tzanelli, 2018). Cinema, lastly, represents a cultural force that creates pseudo-realities which disciplines the non-western Other, I would add, as travel-writing fascinated European readership in the 18th century.

Last but not least, Korstanje and Tzanelli (2019) caution on the risk of consuming (gazing) post-conflict areas. There is a morbid obsession to commoditize the Other's pain while maximizing the pleasure. Based on the West Bank Settlement in Israel, and the demilitarized areas, which are offered the public-emulating anti-terrorist phantasy as the main attraction-, they describe how the needs of being there in would-be insecure place play a leading role in the configuration of dark tourism imaginaries. The scheduled activities range from a sniper tournament to dog attacks or stabbings giving a certain taste of what Israeli army force often face. The ethical dilemma behind these practices is illustrative when authors write:

> *"Fantasy terror camps both obey the logic of domination and act as the apotheosis of consumer capitalism turning the murder to fellow humans into an anodyne spectacle. As part of an infotainment industry specializing in blends of morbidity with patriotic values, terror camps dictate to tourists that they can have fun with impunity. One may argue that contemporary terrorism, tourism, and the value of hospitality are structurally intertwined with histories of European colonization. Historically, Europeans used tourism and the value of hospitality to legitimate expansion, by fashioning civilized travelers-colonists as protectors and educators of the noble savages. Nowadays in globalized contexts of terrorist attacks, such structures of hegemony are reinterpreted, placing the law of hospitality in danger" (Tzanelli and Korstanje, 2019, p. 76).*

Here two assumptions should be done. At a first glimpse, some ethical debate should be situated as a point of entry in this theme and of course, secondly, the introduction of ethics in these types of programs is vital for the individual rights of hosts to be respected. Contrariwise, locals, as well as their cultures, run serious risks to be commoditized. As debate, let me return to the book of Jean and John Comaroff which is seminal in this point. As good anthropologists, the Comaroffs conduct an ethnography in the US and South Africa to understand how cultural industries-like tourism-contribute to the local economies of aboriginal reservoirs. Over the years, these communities have been systematically excluded from the economy or pushed to live in peripheral unproductive lands. As a result of this and even in a democratic country as the US, community's members went through countless risks, and social maladies such as alcoholism, violence, and local crime. During the 90s decade, many communities were seduced by the idea of introducing tourism as a genuine and sustainable form of economic exploitation. Sooner than later, the industry of heritage situated as a promising activity creating new job employment and generating a fairer wealth distribution. When the situation of oppression was finally reversed, the local government imposed higher taxes to these communities instilling a climate of hostility as never before. Although there are many differences between American and South African case, no less true is that tourism may be a vehicle towards a feeling of durable peace, while paradoxically in other conditions, its main obstacle.

5.4   THE AGENDA FOR THE NEXT DECADES

Despite the critical imprint directed against exegetes of tourism as an agent of peace, many examples and books probe tourism generate stable conditions to further peace and cooperation in zones of Rwanda or Northern Ireland. In view of this, the present section revolves around on the challenges and benefits of tourism to regulate conflict. As Thomas Pernecky (2010) put it, most probably because philosophy or ethic was beyond the scope of tourism experts, the philosophical dilemma around peace was widely overlooked. Because of the first studies that focused on tourism and peace came from managerial disciplines, as Pernecky adhere, they undermined the contributions of Martin Heidegger to understand the role of hospitality to reduce the social conflict. The term *being in the World (dasein),* which was proper of Heidegger, sets the pace to *the being of Tourism.* To be more exact, we live in a world where natural beings circulate, for example, men, but this does not mean, these men are tourists since they are born. The notion of what a tourist means is culturally and histori-cally created and internalized. Even it changes according to time. Almost clear concepts as poverty, peace, or injustice, as Pernecky adheres, are defined, and interpreted in different ways by different actors geographi-cally placed in different imaginaries. This suggests that what it is lived is previously conditioned by the biographical background of the observer. A Palestinian Tour guide, naturally, keeps its own idea of geography than a Dutch geography Professor. Their cosmologies are based on contrasting experiences which lead to see the world differently. While traveling, we are being transformed by the "Other" who interrogate us. With basis on this, Kim and Prideaux (2003) decipher the much deeper regional tensions between South and North Koreans. The Hyundai Corporation was legally invited to construct a tourist resort in North Korea. For some reason, South Koreans not only remained indifferent to this commercial establishment but they boycotted it alluding ideological issues. Kim and Prideaux conclude that even if tourism can promote prosperity and peace, in some contexts it instills rivalry and conflict. In this token, Korstanje and George (2012) analyze the case of Malvinas-Falkland, a territory that experienced war in 1982 between the UK and Argentina. Even in dispute today and militarily occupied by the UK, the islands are annually claimed by Argentina. Although Malvinas represents an important cause for the Argentinean imaginary as well as the return of democracy, they avoid visiting Malvinas-Falkland. This does not depend on the hostility of islanders (known as kelpers) who keep the British citizenship, but mass tourism is seen as an industrial machine that objectifies this sacred land. These findings remind not only host-guest interaction is a key factor to undermine or potentiate conflict, but sometimes it is triggered by ideological reasons. For tourism to be successfully adopted, govern-ments should procure a durable peace accompanied by firm and stable institutions (Leslie, 1996). Through participant observation, Clausevic, and Lynch explain that tourism can boost or decline an economy but what is more important some other industries can be benefited. Centered on the case of Bosnia and Herzegovina, they argue that even if there is no substantial evidence that tourism industry gives positive effects to achieve a sustainable integration between divided communities, no less true is that some other industries can indeed do it. Ultimately, Maria Movelli, Nigel Morgan & Carmen Nibigira (2012) lament that civil unrest affects seriously the tourist destination image while declining the economies of poorest nations. Merging the cosmologies of two westerners and one African researcher, they offer a multicultural answer to the problem of conflict in the Third World. Almost 60% of developing countries have experienced a type of civil war or conflict in their territories. In a situation of the fragility of this caliber, tourism would be a valid toolkit to boost local economies. Those countries which have developed substantially in improving their infrastructure have further opportunities in comparison to those which need foreign investments.

5.5 CONCLUSION

Immanuel Kant enthusiastically stressed on the role of hospitality as an organizer of human reciprocity. He envisaged not only an international order, but a league of nations mainly marked by cooperation and coordination. Having said this, Kant alluded to a perpetual pace which is based on mutual understanding situating Europe as a civilizatory order worldwide. In the first section, we explained that tourism research has serious problems to consolidate as a discipline because of the limitations of economic-based paradigm. For this doctrine, tourism is a peaceful instrument of prosperity and development. This chapter discusses critically to what extent tourism promotes peace or conflict, and of course in what conditions it remains as a sustainable industry. There are interesting case-studies that show how tourism-given certain conditions-may very well boost local economies, disarticulating old rivalries, and leading the community towards a prosperous peace. Although tangentially, the problem of terrorism has been marginally approached in this chapter, it is the main theme of the next.

KEYWORDS

- **anthropologists**
- **conflict and violence**
- **global peace index**
- **globalization**
- **hospitality**
- **peace**
- **tourism**

REFERENCES

Barlow, M., (1988). Tourism, peace, and conflict: A geographer's perspective. *Tourism-a Vital Force For Peace, 108.*

Becken, S., & Carmignani, F., (2016). Does tourism lead to peace? *Annals of Tourism Research, 61,* 63–79.

Beirman, D., (2002). Marketing of tourism destinations during a prolonged crisis: Israel and the Middle East. *Journal of Vacation Marketing, 8*(2), 167–176.

Biran, A., & Buda, D. M., (2018). Unravelling fear of death motives in dark tourism. In: Stone, P., Hartmann, R., Seaton, T., Sharpley, R., & White, L., (eds.), *The Palgrave Handbook of Dark Tourism Studies* (pp. 515–532). Palgrave Macmillan, London. Palgrave Macmillan, London.

Blake, A., & Sinclair, M. T., (2003). Tourism crisis management: US response. *Annals of Tourism Research, 30*(4), 813–832.

Centeno, M. A., (2002). *Blood and Debt: War and the Nation-State in Latin America.* Penn State Press, University Park, PE.

Chen, R. J., & Noriega, P., (2004). The impacts of terrorism: Perceptions of faculty and students on safety and security in tourism. *Journal of Travel & Tourism Marketing, 15*(2, 3), 81–97.

Chew, E. Y. T., & Jahari, S. A., (2014). Destination image as a mediator between perceived risks and revisit intention: A case of post-disaster Japan. *Tourism Management, 40,* 382–393.

Cho, M., (2007). A re-examination of tourism and peace: The case of the Mt. Gumgang tourism development on the Korean Peninsula. *Tourism Management, 28*(2), 556–569.

Comaroff, J. L., & Comaroff, J., (2009). *Ethnicity. Inc.* University of Chicago Press, Chicago, IL.

Crouch, G. I., & Ritchie, J. B., (1999). Tourism, competitiveness, and societal prosperity. *Journal of Business Research, 44*(3), 137–152.

D'Amore, L. J., (1988). Tourism—a vital force for peace. *Tourism Management, 9*(2), 151–154.

De Kadt, E., (1979). Tourism: Passport to development. *Perspectives on the Social and Cultural Effects of Tourism in Developing Countries.* World Bank Press; New York, NY.

George, R., (2003). Tourist's perceptions of safety and security while visiting Cape Town. *Tourism Management, 24*(5), 575–585.

Glaesser, D., (2006). *Crisis Management in the Tourism Industry.* London, Routledge.

Hall, C. M., Timothy, D. J., & Duval, D. T., (2004). Security and tourism: Towards a new understanding? *Journal of Travel & Tourism Marketing, 15*(2, 3), 1–18.

Hartmann, R., (2014). Dark tourism, Thana tourism, and dissonance in heritage tourism management: New directions in contemporary tourism research. *Journal of Heritage Tourism, 9*(2), 166–182.

Isaac, R. K., (2010). Alternative tourism: New forms of tourism in Bethlehem for the Palestinian tourism industry. *Current Issues in Tourism, 13*(1), 21–36.

Isaac, R., Cakmak, E., & Butler, R., (2019). Introduction. In: Isaac, R., Cakmak, E., & Butler, R., (eds.), *Tourism and Hospitality in Conflict-Ridden Destinations* (pp. 11–12). Routledge, Abingdon.

Kew, D., & John, A. W. S., (2008). Civil society and peace negotiations: Confronting exclusion. *International Negotiation, 13*(1), 11–36.

Kim, S. S., & Prideaux, B., (2003). Tourism, peace, politics and ideology: Impacts of the Mt. Gumgang tour project in the Korean Peninsula. *Tourism Management, 24*(6), 675–685.

Korstanje, M. E., & George, B. P., (2012). Falklands/Malvinas: A re-examination of the relationship between sacralization and tourism development. *Current Issues in Tourism, 15*(3), 153–165.

Korstanje, M. E., & Tarlow, P., (2012). Being lost: Tourism, risk and vulnerability in the post-'9/11'entertainment industry. *Journal of Tourism and Cultural Change, 10*(1), 22–33.

Korstanje, M. E., (2011). Influence of history in the encounter of guests and hosts. *Anatolia, 22*(2), 282–285.

Laws, E., (2005). *Tourism Crises: Management Responses and Theoretical Insight.* Psychology Press.

Lea, J., (2006). *Tourism and Development in the Third World.* Routledge, Abingdon.

Leslie, D., (1996). Northern Ireland, tourism and peace. *Tourism Management, 17*(1), 51–55.

Litvin, S. W., (1998). Tourism: The world's peace industry? *Journal of Travel Research, 37*(1), 63–66.

Mansfeld, Y., & Pizam, A., (2006). *Tourism, Security and Safety.* Routledge, London.

McKercher, B., & Chon, K., (2004). The over-reaction to SARS and the collapse of Asian tourism. *Annals of Tourism Research, 31*(3), 716–719.

Moufakkir, O., & Kelly, I., (2010). *Tourism, Progress and Peace.* CABI, Wallingford.

Nilsson, D., (2012). Anchoring the peace: Civil society actors in peace accords and durable peace. *International Interactions, 38*(2), 243–266.

Novelli, M., Morgan, N., & Nibigira, C., (2012). Tourism in a post-conflict situation of fragility. *Annals of Tourism Research, 39*(3), 1446–1469.

Pack, S., (2006). *Tourism and Dictatorship: Europe's Peaceful Invasion of Franco's Spain.* Springer, New York, NY.

Patrick, N., (2019). Memorial entrepreneurs and dissonance in post conflict tourism. In: Isaac, R., Cakmak, E., & Butler, R., (eds.), *Tourism and Hospitality in Conflict-Ridden Destinations* (pp. 171–184). Routledge, Abingdon.

Pernecky, T., (2010). The being of tourism. *The Journal of Tourism and Peace Research, 1*(1), 1–15.

Pratt, S., & Liu, A., (2016). Does tourism really lead to peace? A global view. *International Journal of Tourism Research, 18*(1), 82–90.

Richards, G., & Hall, D., (2003). *Tourism and Sustainable Community Development* (Vol. 7). London, Psychology Press.

Ritchie, B. W., (2004). Chaos, crises and disasters: A strategic approach to crisis management in the tourism industry. *Tourism Management, 25*(6), 669–683.

Roberts, C., (2018). Educating the (dark) masses: Dark tourism and sensemaking. In: Stone, P., Hartmann, R., Seaton, T., Sharpley, R., & White, L., (eds.), *The Palgrave Handbook of Dark Tourism Studies* (pp. 603–637). Palgrave Macmillan, London.

Rovelstad, J. M., (1988). World awareness and perception search. *Paper Presented at the First Global Conference: Tourism - a Vital Force for Peace.* Vancouver, Canada.

Salazar, N. B., (2006). *Building a Culture of Peace Through Tourism: Reflexive and Analytical Notes and Queries* (Vol. 62, pp. 319–336). Universitas Humanística.

Salazar, N. B., (2010). *Envisioning Eden: Mobilizing Imaginaries in Tourism and Beyond* (Vol. 31). Oxford, Berghahn Books.

Scott, J., (2012). Tourism, civil society and peace in Cyprus. *Annals of Tourism Research, 39*(4), 2114–2132.

Séraphin, H., Butcher, J., & Korstanje, M., (2017). Challenging the negative images of Haiti at a pre-visit stage using visual online learning materials. *Journal of Policy Research in Tourism, Leisure and Events, 9*(2), 169–181.

Seraphin, H., Korstanje, M., & Gowreesunkar, V. G., (2019). *Post-Disaster and Post-Conflict Tourism: Toward a New Management Approach.* New Jersey, Apple Academic Press.

Suntikul, W., (2019). Tourism and war: Global perspectives. In: Timothy, D., (ed.), *Handbook of Globalization and Tourism* (pp. 139–148). Edward Elgar Publishing, Cheltenham.

Tarlow, P., (2014). *Tourism Security: Strategies for Effectively Managing Travel Risk and Safety.* Oxford, Elsevier.

Timothy, D. J., (2006). Safety and security issues in tourism. In: *Tourism Management Dynamics* (pp. 43–51). Routledge, London.

Timothy, D., (2019). Tourism, border disputes and claims to territorial sovereignty. In: Isaac, R., Cakmak, E., & Butler, R., (eds.), *Tourism and Hospitality in Conflict-Ridden Destinations* (pp. 25–38). Routledge, Abingdon.

Tzanelli, R., & Korstanje, M. E., (2019). 6 On killing the 'toured object'. In: Isaac, R., Cakmak, E., & Butler, R., (eds.), *Tourism and Hospitality in Conflict-Ridden Destinations* (pp. 71–83). Abingdon, Routledge, Abingdon.

Tzanelli, R., (2016). *Thanatourism and Cinematic Representations of Risk: Screening the End of Tourism.* Abingdon, Routledge.

Van Broeck, A. M. (2019). Taking tourism matters into their own hands: Phoenix tourism in Moravia, Medellín, Colombia. In *Tourism and Hospitality in Conflict-Ridden Destinations* (pp. 185-200). Routledge.

Var, T., & Ap, J., (1998). Tourism and world peace. In: Theobald, W., (ed.), *Global Tourism* (pp. 45–47). London, Routledge.

Var, T., Schlüter, R., Ankomah, P., & Lee, T. H., (1989). Tourism and world peace: The case of Argentina. *Annals of Tourism Research, 16*(3), 431–434.

Walters, G., & Mair, J., (2012). The effectiveness of post-disaster recovery marketing messages—The case of the 2009 Australian bushfires. *Journal of Travel & Tourism Marketing, 29*(1), 87–103.

Weaver, A., (2011). Tourism and the military: Pleasure and the war economy. *Annals of Tourism Research, 38*(2), 672–689.

Wise, N. A., (2011). Post-war tourism and the imaginative geographies of Bosnia and Herzegovina, and Croatia. *European Journal of Tourism Research, 4*(1), 5–24.

Yoshida, K., Bui, H. T., & Lee, T. J., (2016). Does tourism illuminate the darkness of Hiroshima and Nagasaki? *Journal of Destination Marketing & Management, 5*(4), 333–340.

CHAPTER 6

Tourism After 9/11: The Day Everything Suddenly Changed

ABSTRACT

It is safe to say that popular opinion over-valorizes tourism as a vehicle towards prosperity and economic health. Once tourism is adopted in local communities, some economic benefits surface. No less true seems to be conflict, violence, and of course, terrorism is serious hazards for the evolution of the tourism industry. Recently, some critical voices have questioned the belief that terrorism affects negatively tourism and other service sectors. Just after the attacks to the World Trade Center as experts argue, tourism experienced rapid recoveries in the different hot-spots or affected destinations. To some extent, terrorism, and tourism are inextricably intertwined. The current chapter inspects the evolution of literature in the fields of tourism security and risk perception theory. Both academic waves, which were centered on the precautionary doctrine, failed to give an all-encompassing diagnosis of the problem.

6.1 INTRODUCTION

On September 11 of 2001, four civilian airplanes were weaponized against the World Trade Center, and the Pentagon affecting not only to the US as a mega superpower, but also the industry of tourism as never before (Levi and Wall, 2004; Bonham, Edmonds, and Mak, 2006; North et al., 2014). This foundational event, which changed the current geopolitical relations between the US and the World, was known as 9/11. At the time to write a book about tourism, one might think about terrorism. Tourism not only was seriously harmed in the post 9/11 days but also the theme was rapidly cast by the media while captivating the attention of numerous scholars worldwide (Korstanje and Tarlow, 2012). Like in 1997 when the horror of Luxor shocked the world, terrorism became in a buzzword term which occupied a central position in conferences, papers, and books. In this chapter, we shall show how tourism security literature has evolved according to the major attacks the US and European countries suffered. To some extent, terrorism, and tourism seem inextricably intertwined.

As the previous argument is given, theorists have recently interrogated the concept of mobilities. Being mobile exhibits a new sign of status which marks a privileged (capitalist) class from others which are not encouraged to travel and visit other countries (Sheller and Urry, 2006; Cresswell, 2016). Because of this, it is not surprising that terrorists selected commercial airplanes to humiliate to one of the most powerful nations in the world (Korstanje, 2018). This chapter aims to give fresh answers to understand the reasons why modern tourists are targeted by radicalized groups, a tendency that came to stay just after 9/11 (Saha and Gap, 2014; Tzanelli, 2018; Korstanje and Clayton, 2012). In a seminal book, *The Political economy of terrorism,* Enders*, and* Sandler calls attention to the complex interplay between terrorism and tourism. Per their viewpoint, terrorists look to maximize their gains while reducing costs. Terrorist behavior can be very well explained through the rationality of the economy. They hold the thesis that because of its nature, tourism, and leisure-spots are spaces of freedom and mobility where security forces do not intrude. Secondly and most important, lay-people panicked when the attacks happen in public or commonplaces. This begs a more than interesting point: is tourism receding towards the advance of international terrorism or is terrorism an instrument to expand tourism?

Experts do not agree about these questions. For that reason, the present chapter intends to fill the gap. The first section deals with a preliminary debate on the theme stressing for the main contributions in the field. Secondly, we focus on the borrowing of risk perception theory, a theory that gained a reputation in social sciences over four decades, in tourism research. Among the limitations of risk perception theory, we enumerate the proliferation of quantitative-related methods (over qualitative ones) adjoined to the lack of a historical perspective to understand the nature of terrorism or at least how terrorist attacks molded the specialized literature. This means that we understand this much deeper issue through the lens preconceptions, ideas, or fears.

6.2   PRELIMINARY DEBATE

One of the frightful aspects of terrorism are based not only in the sudden factor but in the possibilities to target (victimize) innocent people who are not involved directly in the battleground. As McConaghy puts it, the attacks to non-combatants often violate the laws of war. Non-combatants are not considered direct targets in the battleground but as some experts agree, there are collateral damages. The paradox of terrorism may be very well explained following this axiom: though the considerable financial investment of governmental agencies to enhance security, less is known of terrorism as an object of study. Governments react differently to terrorist attacks. Sometimes, governments adopt populist discourses that distance them from a coherent diagnosis of the problem. Terrorism not only harmed tourism as specialists known, but also is gradually changing our cultural entertainment industries (Spigel, 2004; Altheide, 2004; Sthal, 2009; Jackson, 2011; Korstanje, 2018). The point was brilliantly investigated by Darcie Rives-East in her book Surveillance and Terror in Post-9/11 British and American Television. After 9/11, the social imaginary in the US and the UK felt captivated to consume stories of criminality, torture, and imprisonment. These documentaries activated long-dormant discourses incipient in the American-character formation. The narratives of captivity as well as the ideals of exception, which was proper of Puritanism, played a leading role to portray "the non-Western Other." Far from disappearing, these discourses are present in the counter-terrorism culture. An industry of electronic surveillance accompanied by a sentiment anti-immigration policy was some of the remarkable signs of the post 9/11 days. The culture of surveillance and homeland security accelerated a rapid growth of these agencies in the name of a safer society, but at the same time, it compromises the founding cultural values of the US and the UK, as the author alerts (Rives-East, 2019). Is tourism changing to the advance of terrorism?

One of the pioneering voices who focused her attention on the theme of terrorism in the tourism fields, was Sevil Somnez, who in 1998 argued those countries which often fail to keep terrorism under control is widely characterized by durable political crises which led them to develop unstable institutions. Her position associates to the needs of understanding the issue to protect the tourism industry. The problem given in these terms seems to be that under-developing economies have a dependent situation respecting to tourism (Somnez, 1998). In the former chapter, we have discussed the position of economic-based paradigm to punctuate that consolidated democracy has further opportunities to develop sustainable tourism. The same applies to the case of terrorism. In this respect, Somnez echoes an academic doctrine that overtly acknowledges that democracy acts as an antidote to mitigate the radicalized ideologies and the economic discontent resulted from the countless economic crises in the Middle East. For these voices, the western rationality and the planning process suffice to contain the risk of terrorism-at the least by the articulation of communicative campaigns oriented to mitigate the negative effects of terrorism in local economies (Somnez, 1998; Somnez and Graefe, 1998; Somnez, Apostolopoulos, and Tarlow, 1999; Ritchie, 2004; Hall, Timothy, and Duval, 2012). The main goals of terrorists linked to the decline of social ties, as well as creating panic to negotiate with the government, as Richter and Waugh (1986) claim. Because of this, and of course the fragility of tourism as a commercial activity, some radicalized groups select tourist areas to captivate media attention.

The main point of entry stems from the unilateral assumption that terrorists are resented peoples who hate democratic and capitalist countries. Western tourists are being targeted because of the serious material inequalities between the Global North and the South. In this vein, terrorism operates in clandestine spheres, probably located in undemocratic and undeveloped countries (Mansfeld, 1999; Somnez and Graefe, 1998; McKercher and Hui, 2004). Let us clarify that the economic-based paradigm has prioritized the spill-over effect as the main variable to test, which suggests that terrorist attacks affect the image of neighboring destinations. For example, Pizam and Fleisher (2002) found that the frequency of terrorist attacks seriously affects the destination image, accelerating its decline in the threshold of time. In this vein, these researchers validate the hypothesis that frequency-rather than severity-is the most negative campaign for a tourist destination. Here one must distinguish the political violence exerted by terrorism from the local crime. Even though both risks place tourists in jeopardy, they are different. It is noteworthy that tourists are unfamiliar with the visited place, as well as they are identified from the local population. As Chrys Ryan puts it, tourism particularly is an activity that attracts criminals. One of the motivations of thieves or criminals rests on the economic-factor. Tourists who come from First world countries- and dotted with high purchasing power-have more probabilities to be victimized than other segments. Instead, terrorists are psychologically motivated by ideological and political aims. Tourists are seen as agents of capitalism who would be punished. State-sponsored tourism is the tug of war of some underdeveloped nations, for this reason, an attack to tourism represents an attack to the government. However, despite its credibility, this conceptual position was radically shifted after the attacks to the World Trade Center in 2001. This bloody event not only has drawn the attention of the world, but tourism-related researchers have also seen in terrorism and political violence as a major threat to defeat. Even if the risk perception theory had more than four decades in the field of social sciences, it was not widely cited in tourism research. In the post 9/11 contexts, tourism scholars were enthused to adopt risk perception theory as a guiding conceptual platform to follow (Fuchs and Reichel, 2006; Yang and Nair, 2014; Jonas et al., 2011; Yong, Khoo-Lattimore, and Arcodia, 2017). To the problem of terrorism, many other risks added as natural disasters, virus outbreaks only to name a few (Reisinger and Mavondo, 2005; Waugh and Smith, 2006; Pezzulo, 2009; Smith, Speers, and MacKenzie, 2011). In the next section, we shall explore the contribution and limitations of risk perception theory exclusively applied to terrorism.

6.3   RISK PERCEPTION, TOURISM, AND TERRORISM

Over the years, scholars have questioned the efficiency of western government to struggle against radicalized jihadism. As introduced in the earlier section, a group of scholars adopted risk perception theory to understand not only the evolution of fear but also how destinations are finally perceived (Kozak, Crotts, and Law, 2007; Kutto and Grooves, 2004). Moved by the needs of understanding the impact of risk perception for the market, Wesley Roehl and Daniel Fesenmaier (1992) conduct applied research to understand how risk perception is cognitively structured. Their investigation ultimately concludes that three factors are important to explain why some destinations are preferred or avoided: demographic variables of travelers, the type of travel (business-pleasure travel) and situation-specific factors. What this investigation reveals are that the role of travelers plays an important role in how risks are imagined and perceived. The typology offered by Roehl and Fesenmaier (1992) contributed notably to the tourism research and was widely cited in countless studies (Somnez and Graefe, 1998; Reynolds and Braithwaite, 2001; Lepp and Gibson,

2003; Gursoy and McCleary, 2004; Reisinger and Mavondo, 2005; Munar and Jacobsen, 2014). Motivated by this chapter, in 2004, Myron et al. publish a seminal paper which entitles *the effects of risk perception on intention to travel in the aftermaths of September 11, 2001*. The research sees the light of publicity in the prestigious journal: *Journal of Travel and Tourism Marketing*. In this text, researchers want to expand the current understanding between the perception of New Yorkers in the post 9/11 contexts, and their intentions to travel abroad. The main findings reveal-as authors assume-two important points. On the one hand, terrorism creates negative effects in New Yorkers for a lapse of one year (or near circa). Secondly, those leisure travelers who planned to take a trip abroad were more prone to safety concern than business-led travelers. Terrorism and 9/11 in particular not only instilled panic in the society but also canceled thousands of trips and hotel accommodation noting millions in losses for the hospitality and tourism industries. This coincides with the remarks of Lepp and Gibson (2003) who argue convincingly that the act of traveling (abroad) involves two contrasting forces. Visitors need to feel secure at the destination, at the same time they are in quest of novel experiences-as we have explained in the chapters that define tourism as a rite of passage. In this case, tourists develop different degree to tolerate uncertainty leading to countless capacities and reactions to face risks. This is in confidence with the study of Reisinger and Mavondo (2005) who establish that each personality adapts differently to frustration and uncertainty. Consumers made their decisions combining numerous emotional and cognitive frames. This explains why risks are personally negotiated and perceived.

To cut the long story short, the following scheme summarizes the obtained results revolving around the influence on risks in the decision-making process:

- Demographic variables such as identity, religion, territorial belonging, nationality, or age, are directly proportional to risk perception (Woods et al., 2008; Lepp and Gibson, 2003; Ertuna and Ertuna, 2009).
- Business travelers are not psychologically sensitive to visit risky destinations than pleasure travelers (Park and Reisinger, 2010; Floyd et al., 2004).
- The role of travelers, as well as their disposition to bear uncertainty plays a leading role in the formation of risks (Reichel, Fuchs, and Uriely, 2007; Reisinger and Mavondo, 2005).

- Some previous traumatic experiences potentiate a sentiment of avoidance of some destination. Events, which are mediatically broadcasted and massively disseminated, are memorized in the social imaginary in a threshold of time (Gut and Jarell, 2010).
- After 9/11, urban destinations were avoided while the demand preferred rural zones or countryside (Floyd et al., 2004; Yuan, 2005).

In a nutshell, in what has been considered as the bloodiest attack on American soil after Pearl Harbor, September 11 represented a turning point for specialists and policy-makers in the world (Dalby, 2003). From that moment onwards, the world was cut in two sides: secure and insecure nations. What is more important, 9/11 marked the first-time terrorists employed mass means of transport as real weapons against civilian targets (Korstanje and Olsen, 2011). Since 9/11, countless studies focused on the economic effects of terrorism in local economies, as well as tourism and hospitality industries (Bonham, Edmons, and Mak, 2006; Pappas, 2010; Korstanje and Clayton, 2012; Saha and Yap, 2014; Raine, 2013; Yan et al., 2016). The risk perception model gave to researchers a coherent model to be applied in their multicultural scopes. Although the theory of risk perception pivoted in the construction of an all-encompassing model that predicts terrorist attacks in leisure hot-spots (Reisinger and Mavondo, 2005; Floyd et al., 2004; Fuchs et al., 2013; no less true was that the dilemma of security opened the doors to new unanswered questions.

At a first glimpse, Professor Larsen and his team showed recently that risk perception is formed in the cognitive system which suggests it is associated with the information the subject receives. This means that people may perceive a destination as unsafe while they wish to travel there in any way. Under some conditions, risk, and emotionality are dissociated. What is more important, while some risks are exaggerated by the subject others are systematically ignored (Larsen, 2007; Larsen, Brun, and Ogaard, 2009; Wolff and Larsen, 2014). Far from being rational agents, tourists are motivated by internal emotions, which connect to the previous experiences. Because of this, the figure of fear surfaces as an explanatory variable that helps policy-makers to articulate the necessary steps towards more secure destinations (Adams, 2001; Bianchi, 2006). Secondly, the interests on what tourist feel or how they perceive the international destinations led researchers to an ecological fallacy, getting contradictory evidence that put the discipline in a stagnant point. A great proportion of the studies emphasized on the statistical instrument or quantitative methodologies oriented to describe how destinations were perceived. However, under some conditions people lie to protect their interests or simply are unfamiliar with their inner-world. This begs some pungent (methodological) questions, to what an extent does a tourist help scholars in understanding terrorism? or what is worse, is terrorism what tourists feel it is?

In sum, tourism security researchers echoed the risk perception theory as the main paradigm, but in so doing, they failed prey of their stereotypes and Euro-centrism that assumes there are secure p and insecure places. At the same time, we feel we are living safer, invariably we believe that others dwell in unsafe conditions. As Raoul Bianchi (2006) puts it, the excessive attention paid by public opinion on the features of the destinations as well as its capabilities to strengthen security has invariably led towards a problematic position. The fear of terrorism, adjoined to the globalization of anxiety, paved the pathways for the rise of a new climate where despite the efforts and material resources to make a destination safer, the public are entangled in a state of emotional vulnerability which reinforces the logic of terrorism (Bianchi, 2006; Bianchi and Stephenson, 2013).

Last but not least, Korstanje and Clayton (2012) ignite a hot debate revolving the nature of terrorism and its connection in tourism. Far from harming the activity of tourism industry, these experts say, tourism, and terrorism share many commonalities. This thesis starts from the premise that the western rationality introduces extortion as a form of relationship. Terrorists look to get further gains (generating a shock in the global spectatorship) by minimizing their costs. In the cycles of capitalism, which is associated with a logic of destructive creation, terrorism allows the capital circulation in ways otherwise would not be feasible. As a result of this, the ongoing atmosphere of fear stimulates mass-consumerism but at a higher cost. For the sake of clarity, security situates as a commodity which is exchanged in different economic circuits, enlarging the gap between central and peripheral economies.

The methodological limitations of risk perception theory have been emphasized in the earlier chapters. Among them we may cite the lack of qualitative methods to understand the nature of fear and risk perception, the emphasis on what the tourist feels like the only methodological source of information, the lack of a historical insight that helps understand terrorism (only to name a few). Next, we will review the three stages that evolved tourism security from its inception to date.

6.4 THE EVOLUTION OF LITERATURE IN TOURISM SAFETY AND SECURITY

As debated, the model of tourism-management, which prioritized concrete policies to protect the tourist destinations, was mainly oriented to control "the spillover effects." This means that the negative aftermaths of terrorism straddle the affected country, on some occasions involving the neighboring areas or nations (Drakos and Kuttan, 2003). To put an example even if some nations-geographically located in the Middle East-are free of terrorist activity, the Middle East is often pondered as an insecure region for international demand (Pizam and Fleisher, 2002; Bassil, 2014). However, at a closer look, we shall see that each group of theories, which was focusing on tourism security, was historically inscribed in a specific-based context.

The inception of the discipline is widely marked by the Luxor Massacre, a terrorist attacks perpetrated in Egypt by the group Al-Jama'a al-Islamiyya where 62 foreign tourists are assassinated. The attack takes place on November 17 of 1997 at the mortuary Temple of Hatshepsut. Armed with automatic weapons six gunmen kidnapped a tourist contingent. A tourist is systematically killed every 45 minutes. The youngest victim was a five-year-old British child. This event not only shocked the Western imaginary, placing Egypt as an extremely dangerous destination, but also inspired to some theorists to turn their attention to the *scourge of terrorism*. In her respective publications, Sevil Somnez (1998) alert on the problem of terrorism as a major threat for the tourism industry and the political stability of the Middle East. The tourism industry when it is dully regulated acts as a mechanism that boosts local economies, through the multiplying factor or creating the conditions towards a fairer wealth distribution. Here the problem lays in the fact some countries where the political conditions are not granted, adopt tourism as their main economic relief. Those countries castigated by civil wars or undemocratic institutions not only fail to develop tourism but also runs serious risks respecting to terrorism. The dependency of central administration to the economic benefits of the tourism industry, as well as the rise of the radicalized group, lays the foundations towards a chaotic situation very hard to reverse. As Somnez bemoans, some radicalized groups target tourists to accelerate a climate of political instability affecting not only the profits of the community but the organic image of the country. Like Somnez, those researchers, who have investigated on tourism security, are profoundly worried by the immediate consequences of terrorism and political violence in the tourism, hospitality, and service sectors. This is exactly the case of Abraham Pizam who is a leading voice in these types of issues. In Pizam's approach, at the time scholars think the fields of tourism security, it is important not to lose the sight of the fact the violence perpetrated by terrorism is not the only factor. Acts of domestic violence, accompanied by local crime are key elements to be taken seriously into consideration (Mansfeld and Pizam, 2006; Pizam, 1999; Fuchs and Pizam, 2011).

With the benefit of hindsight, the second event that molded the discipline was the attack to the World trade center occurred in 2001 (as debated in earlier sections of this chapter). Such an event was radically different than the Luxor Massacre. To the moment, scholars believed that democratic countries were immune to the act of terrorism. The 9/11 showed not only the opposite but how democratic nations geographically in the Global North may be very well targeted by terrorist cells. The adoption of risk perception theory seems to be aligned to the needs of implementing an urgent solution that protects the tourism industry. Peter Tarlow's works are highly influenced by this idea there is not a safe place any longer, anyone, and anywhere can be a victim of terrorism. As he eloquently puts it, the utopian-if not perfect-a vision that western rationality may reach a risk-zero society blurs with the urgency to find a coherent solution for the obstacles tourism security is going through. In a hyper-mobile society, immersed in a climate of complexity and global risks, the tourism industry is being placed in jeopardy. No less true is that some voices go further and event alert on the end of tourism-at least as we know it (Hannam, 2009; Gale, 2009).

It is safe to say that the third even that changed the conception of tourism security was the recent attacks on Paris, Brussels, and London since 2015 by the Islamic State. This was considered a radical turning point-even for promoters of tourism security-because it probed those terrorists were not foreigners or people who arrived in Europe to orchestrate the attack, they rather were born-native from the societies they hated. Those studies emerged in this period captivated the negative effects on terrorism over the tourism industry (like other works) but went beyond, alerting on the decline of hospitality. Terrorism is recreating a state of paranoia that led West to close borders, while sees the Non-Western Other as a potential terrorist. Centered on the idea of the enemy living within, this intolerance to accept the *Otherness,* gradually leads the Western civilization to the erosion of their democratic institutions, even the law of hospitality which was histori-cally its mainstream value. In other studies (Korstanje and Tarlow, 2012; Korstanje and Olsen, 2011), we analyzed dozens of horror movies made from 2001 to date. We found that terrorism is not only affecting how the "Other" is perceived but also how hospitality is culturally being imagined and re-negotiated. The plots of these movies start with young (white) tour-ists who often come from the US or Europe. They opt to travel-probably as backpackers-towards exotic places geographically located in Eastern Europe or Asia. These naïve travelers are targeted by monsters, or psycho-paths who loom from the darkness. They are invited to eat or drink and finally kidnapped while sleeping. Not surprisingly, the enemy (internal risk) is imagined as living here, (not there) looking and behaving like us. In view of the contemporary events, it is safe to say that Donald Trump reached the US presidency promising the construction of a Wall between the US and Mexico border. While a globalized process has flattened the earth, no less true seems to be that a counter-force which is based on ethnocentrism and racism, recently surfaced. These nationalist discourses are oriented to demonize the "foreigner" as an undesired "Other." The question of whether hospitality consists of accepting the Other as a part of society, echoing Selwyn (2019), a much deeper sentiment of hostility has been empowered in Europe and the US. This suggests that there are two types of mobilities. One strictly reserved for rich tourists who are legally enthralled to travel across the world, while the other signals to the surveil-lance and controls over forced-migrants, asylum seekers and refugees. The triumph of Trump appears not to be different than Jair Bolsonaro or Viktor Orban-without mentioning the Brexit and other separatist movements in Europe. In the past, the Western empires adventured to colonize the world. In so doing, they imposed the western hospitality as a mainstream cultural value. Today, the law of hospitality sets the pace to a resistance to the "non-Western Other." The rise of neo-fascism, as well as Islamophobia-if not the tourist-phobia, places the problem of racism into the foreground. As David Altheide puts it, the sentiment of terror generated by terrorism not only allowed the emergence of long-dormant discourses programed to revitalize the most recalcitrant chauvinism but also affected the democratic institutions as never before. The legitimacy of Trump, as Altheide notes, is not given by the fear of strangers, which was enrooted in the American character, but in the urgency of enhancing homeland security (Altheide, 2006, 2017). Equally important, terrorism not only shocked the US and the World but also reminded that anyone and anytime can be a victim. This idea woke up a sentiment of living with the enemy within (Ansari, 2018), which is the symbolic core of Islamophobia (Korstanje, 2018). As Selwyn (2019) brilliantly writes, England in particular and Europe, in general, are coming across a sentiment of anti-hospitality, which consists in expulsing some ethnic minorities that took part of society over decades to the borders of the system (Windrush Scandal). At the bottom, hospitality reflects human reciprocity exhibiting a logic of giving-while-receiving. Such solidarity forms the social institutions and the norms the in-group members often obey. However, one might speculate that the term is strictly associated with hostility. The acts of hospitality denote the articulation of durable social relationships giving symbolic frames to move inside or outside groups once consolidated. One of the anthropological functions of hospitality is to transform enemies in friends, strangers in neighbors. A hostile environment, broadly speaking, does the opposite, the neighbor who may be suspected of being a potential terrorist-is monitored, jailed, or deported with some reasonable argument. In view of this, as Selwyn observes, Brexit, and Windrush Scandal are two sides of the same coin for the UK. Nevertheless, this should not be limited to England. General observations about anti-tourist manifestations or anti-Islam parades speak readers of the rise and expanse of an anti-hospitality discourse.

6.5   CONCLUSION

Tourism Security and safety are vital to the well-functioning of tourism. Over recent years, to be more exact, since 2001, the tourist system was whipped by numerous risks such as virus outbreaks, natural disasters, and terrorism. This chapter discussed the evolution of tourism safety literature according to how terrorism impacted society. Undoubtedly, the Luxor Massacre, 9/11 or November 2015 Paris attacks marked our lives and the ways travels are perceived. We have reviewed the most significant studies that form what academicians know as *Risk Perception theory,* while we lay a critical foundation of its limitations. The chapter reminds the importance to unearth a historical insight to understand not only the evolution of terrorism but how-epistemologically speaking-tourism safety theory is shaped. Today risk perception theory lacks a robust methodological background that invariably led to a conceptual gridlock.

KEYWORDS

- **conflict**
- **methodologies**
- **mobilities**
- **political instability**
- **risk perception theory**
- **terrorism**
- **war**

REFERENCES

Adams, K. M., (2001). Danger-zone tourism: Prospects and problems for tourism in tumultuous times. In: *Interconnected Worlds* (pp. 265–281). Elsevier, Oxford.

Altheide, D. L., (2004). Consuming terrorism. *Symbolic Interaction, 27*(3), 289–308.

Altheide, D. L., (2006). "Terrorism and the politics of fear. Cultural studies?" *Critical Methodologies, 6*(4), 415–439.

Altheide, D. L., (2017). *Creating Fear: News and the Construction of Crisis*. Routledge, Abingdon.

Ansari, H., (2018). *"The Infidel Within": Muslims in Britain Since 1800*. Oxford University Press, Oxford.

Bassil, C., (2014). The effect of terrorism on tourism demand in the Middle East. *Peace Economics, Peace Science and Public Policy, 20*(4), 669–684.

Bianchi, R. V., & Stephenson, M. L., (2013). Deciphering tourism and citizenship in a globalized world. *Tourism Management, 39*, 10–20.

Bianchi, R., (2006). Tourism and the globalization of fear: Analyzing the politics of risk and (in) security in global travel. *Tourism and Hospitality Research, 7*(1), 64–74.

Bonham, C., Edmonds, C., & Mak, J., (2006). The impact of 9/11 and other terrible global events on tourism in the United States and Hawaii. *Journal of Travel Research, 45*(1), 99–110.

Cresswell, T., (2016). Towards a politics of mobility. In: *Routes, Roads and Landscapes* (pp. 181–196). Routledge, Abingdon.

Dalby, S., (2003). Calling 911: Geopolitics, security and America's new war. *Geopolitics, 8*(3), 61–86.

Drakos, K., & Kutan, A. M., (2003). Regional effects of terrorism on tourism in three Mediterranean countries. *Journal of Conflict Resolution, 47*(5), 621–641.

Enders, W., & Sandler, T., (2011). *The Political Economy of Terrorism*. Cambridge, Cambridge University Press.

Ertuna, C., & Ertuna, Z. I., (2009). The sensitivity of German and British tourists to new shocks. *Tourism Review, 64*(3), 19–27.

Floyd, M. F., Gibson, H., Pennington-Gray, L., & Thapa, B., (2004). The effect of risk perceptions on intentions to travel in the aftermath of September 11, 2001. *Journal of Travel & Tourism Marketing, 15*(2, 3), 19–38.

Fuchs, G., & Pizam, A., (2011). 18 the importance of safety and security for tourism destinations. In: Wang, Y., & Pizam, A., (eds.), *Destination Marketing and Management* (pp. 300–312). Wallingford, Cabi.

Fuchs, G., & Reichel, A., (2006). Tourist destination risk perception: The case of Israel. *Journal of Hospitality & Leisure Marketing, 14*(2), 83–108.

Gale, T., (2009). Urban beaches, virtual worlds and 'the end of tourism'. *Mobilities, 4*(1), 119–138.

Gursoy, D., & McCleary, K. W., (2004). An integrative model of tourist information search behavior. *Annals of Tourism Research, 31*(2), 353–373.

Gut, P., & Jarrell, S., (2010). Silver lining on a dark cloud: The impact of 9/11 on a regional tourist destination. *Journal of Travel Research, 46*, 147–153.

Hall, C. M., Timothy, D. J., & Duval, D. T., (2012). *Safety and Security in Tourism: Relationships, Management, and Marketing*. Routledge, Abingdon.

Hannam, K., (2009). The end of tourism? Nomadology and the mobilities paradigm. In: Tribe, J., (ed.), *Philosophical Issues in Tourism, 37*, 101. Bristol, Channel View Publications.

Jackson, R., (2011). Culture, identity and hegemony: Continuity and (the lack of) change in US counterterrorism policy from Bush to Obama. *International Politics, 48*(2, 3), 390–411.

Jonas, A., Mansfeld, Y., Paz, S., & Potasman, I., (2011). Determinants of health risk perception among low-risk-taking tourists traveling to developing countries. *Journal of Travel Research, 50*(1), 87–99.

Korstanje, M. E., & Clayton, A., (2012). Tourism and terrorism: Conflicts and commonalities. *Worldwide Hospitality and Tourism Themes, 4*(1), 8–25.

Korstanje, M. E., & Tarlow, P., (2012). Being lost: Tourism, risk and vulnerability in the post-'9/11'entertainment industry. *Journal of Tourism and Cultural Change, 10*(1), 22–33.

Korstanje, M. E., (2018). *The Challenges of Democracy in the War on Terror: The Liberal State Before the Advance of Terrorism*. Routledge, Abingdon.

Kozak, M., Crotts, J., & Law, R., (2007). The Impact of the perception of risk on international travelers. *International Journal of Tourism Research, 9*(4), 233–242.

Kuto, B., & Groves, J., (2004). "The effects of terrorism: Evaluating Kenya's tourism Crisis. *E-Review of tourism Research, 2*(4), 88–95.

Lepp, A., & Gibson, H., (2003). Tourist roles, perceived risk and international tourism. *Annals of Tourism Research, 30*(3), 606–624.

Levi, M., & Wall, D. S., (2004). Technologies, security, and privacy in the post 9/11 European information society. *Journal of Law and Society, 31*(2), 194–220.

Mansfeld, Y., & Pizam, A., (2006). *Tourism, Security and Safety*. Abingdon, Routledge.

McConaghy, K., (2017). *Terrorism and the State: Intra-State Dynamics and the Response to Non-State Political Violence*. Palgrave-MacMillan, New York, NY.

McKercher, B., & Hui, E. L., (2004). Terrorism, economic uncertainty and outbound travel from Hong Kong. *Journal of Travel & Tourism Marketing, 15*(2, 3), 99–115.

Munar, A. M., & Jacobsen, J. K. S., (2014). Motivations for sharing tourism experiences through social media. *Tourism Management, 43*, 46–54.

North, C. S., Gordon, M., Kim, Y. S., Wallace, N. E., Smith, R. P., Pfefferbaum, B., & Pollio, D. E., (2014). Expression of ethnic prejudice in focus groups from agencies affected by the 9/11 attacks on the world trade center. *Journal of Ethnic and Cultural Diversity in Social Work, 23*(2), 93–109.

Pappas, N., (2010). Terrorism and tourism: The way travelers select airlines and destinations. *Journal of Air Transport Studies, 1*(2), 76–96.

Park, K., & Reisinger, Y., (2010). Differences in the perceived influence of natural disasters and travel risk on international travel. *Tourism Geographies, 12*(1), 1–24.

Pezzullo, P. C., (2009). "This is the only tour that sells": Tourism, disaster, and national identity in New Orleans. *Journal of Tourism and Cultural Change, 7*(2), 99–114.

Pizam, A., & Fleischer, A., (2002). Severity versus frequency of acts of terrorism: Which has a larger impact on tourism demand? *Journal of Travel Research, 40*(3), 337–339.

Pizam, A., (1999). A comprehensive approach to classifying acts of crime and violence at tourism destinations. *Journal of Travel Research, 38*(1), 5–12.

Raine, R., (2013). A dark tourist spectrum. *International Journal of Culture, Tourism and Hospitality Research, 7*(3), 242–256.

Reichel, A., Fuchs, G., & Uriely, N., (2007). Perceived risk and the non-institutionalized tourist role: The case of Israeli student ex-backpackers. *Journal of Travel Research, 46*(2), 217–226.

Reisinger, Y., & Mavondo, F., (2005). Travel anxiety and intentions to travel internationally: Implications of travel risk perception. *Journal of Travel Research, 43*(3), 212–225.

Richter, L. K., & Waugh, Jr. W. L., (1986). Terrorism and tourism as logical companions. *Tourism Management, 7*(4), 230–238.

Ritchie, B. W., (2004). Chaos, crises and disasters: A strategic approach to crisis management in the tourism industry. *Tourism Management, 25*(6), 669–683.

Rives-East, D., (2019). *Surveillance and Terror in Post-9/11 British and American Television.* Darcie Rives-East. Palgrave Macmillan, New York, NY.

Roehl, W. S., & Fesenmaier, D. R., (1992). Risk perceptions and pleasure travel: An exploratory analysis. *Journal of Travel Research, 30*(4), 17–26.

Ryan, C., (1993). Crime, violence, terrorism and tourism: An accidental or intrinsic relationship? *Tourism Management, 14*(3), 173–183.

Saha, S., & Yap, G., (2014). The moderation effects of political instability and terrorism on tourism development: A cross-country panel analysis. *Journal of Travel Research, 53*(4), 509–521.

Selwyn, T., (2019). Hostility and hospitality: Connecting Brexit, Grenfell and Windrush. In: Rowson, B., & Lashley, C., (eds.), *Experiencing Hospitality.* New York, Nova Science publishers.

Sheller, M., & Urry, J., (2004). *Tourism Mobilities: Places to Play, Places in Play.* Routledge, London.

Smith, D. W., Speers, D. J., & Mackenzie, J. S., (2011). The viruses of Australia and the risk to tourists. *Travel Medicine and Infectious Disease, 9*(3), 113–125.

Sönmez, S. F., & Graefe, A. R., (1998). Influence of terrorism risk on foreign tourism decisions. *Annals of Tourism Research, 25*(1), 112–144.

Sönmez, S. F., (1998). Tourism, terrorism, and political instability. *Annals of Tourism Research, 25*(2), 416–456.

Sönmez, S. F., Apostolopoulos, Y., & Tarlow, P., (1999). Tourism in crisis: Managing the effects of terrorism. *Journal of Travel Research, 38*(1), 13–18.

Spigel, L., (2004). Entertainment wars: Television culture after 9/11. *American Quarterly, 56*(2), 235–270.

Stahl, R., (2009). *Militainment, Inc.: War, Media, and Popular Culture*. Routledge, London.

Tzanelli, R., (2018). *Mega Events as Economies of Imagination: Creating Atmospheres for Rio 2016 and Tokyo 2020*. Routledge, Abingdon.

Waugh, Jr. W. L., & Smith, R. B., (2006). Economic development and reconstruction on the Gulf after Katrina. *Economic Development Quarterly, 20*(3), 211–218.

Woods, J., Eyck, T. A. T., Kaplowitz, S. A., & Shlapentokh, V., (2008). Terrorism risk perceptions and proximity to primary terrorist targets: How close is too close? *Human Ecology Review*, 63–70.

Yan, B. J., Zhang, J., Zhang, H. L., Lu, S. J., & Guo, Y. R., (2016). Investigating the motivation-experience relationship in a dark tourism space: A case study of the Beichuan earthquake relics, China. *Tourism Management, 53*, 108–121.

Yang, C. L., & Nair, V., (2014). Risk perception study in tourism: Are we really measuring perceived risk. *Procedia-Social and Behavioral Sciences, 144*(1), 322–327.

Yang, E. C. L., Khoo-Lattimore, C., & Arcodia, C., (2017). A systematic literature review of risk and gender research in tourism. *Tourism Management, 58*, 89–100.

Yuan, M., (2005). After September 11: Determining its impacts on rural Canadians travel to U.S. *E-Review of Tourism Research, 3*(5), 103–108.

CHAPTER 7

Is Hospitality Dying? In Robots We Must Not Trust

ABSTRACT

The occident's imaginary the relation of self and alterity was at least troublesome. Originated in the core of colonial rule to dominate what European minds considered the savages, the dichotomy between rationality and emotionality served to forge a dominant discourse that widely legitimated the conquest and annexation of over-seas territories (colonies). Today's tourism faces one of its more complex dilemmas in view of the rise of a new virus outbreak known as COVID-19. With the world strongly shocked and paralyzed, scholars interrogate furtherly about the future of tourism in the years to come. Some voices have claimed that this is the time to introduce robots and IA in the services of guests marking a serious turning point in the ways tourism was practiced. The preliminary chapters of this book discussed-from different angles-the nature and evolution of tourism. However, nowadays the convergence between global threats and virtual technologies leads us to re-think that the Sacred laws of hospitality, as they were forged in the Western civilization, are in decline. Are we experiencing the ends of tourism as we know it? Or simply it relates to the rise of new forms of unethical forms of consumption? Is tourism mutating towards more morbid gazes?

7.1 INTRODUCTION

The latest virus outbreak that has shocked the world (COVID-19) in the Chinese province of Wuhan poses as a significant threat to the tourism industry today. It is safe to say this virus represents one of the major risks the sector faced in its history-without mentioning WWII-. In view of the problems had European countries to stop the Pandemic, the world closed its internal borders. This decision not only affected the industry of tourism and hospitality but other service sectors. Authorities and policy-makers estimate that the material losses will be in millions. In the mid of this mayhem, some voices have asked for the introduction of robots and artificial intelligence (AI) to deal with clients and guests. Quite aside from the impossibility of this step, the idea speaks to us of what some scholars have dubbed as the End of Hospitality (Korstanje, 2017; Innerarity, 2001) or the end of tourism (Gale, 2009). Starting from the premise that Coronavirus will be a turning point in the history of Tourism, the present chapter is fully dedicated to the role of hospitality in the configuration of Western social imaginary. The West has been based on the *Sacred Law of hospitality* (which comes from the Greek Spirit). Today this symbolic touchstone seems to be threatened by countless risks such as terrorism, white supremacists, Islamophobia, and of course, COVID-19. The change of millennium has brought radical shifts, fears, and global risks, for which the industry was not prepared. Of course, the problem of global risks is not the original object of study in this chapter, because the point was hotly debated in other sections. The current chapter is reserved to *the end of hospitality* as a common-thread argument. We analyze the plot of the film *The Passengers* to theorize on the complex interlink between hospitality and robots today. Although some parts follow philosophical debates that must be hard to follow for people educated in other disciplines, we feel the results deserve our time. We combine the advances in the anthropology of hospitality and the philosophical dilemmas revolving around the sense of reality. Ultimately, the plot of Passengers gives readers a snapshot of what we dubbed as a *failed hospitality.*

7.2   THE NATURE OF HOSPITALITY

The word hospitality stems from *Hospitium (lat),* a term which was employed in ancient Rome to regulate the flux of foreign citizens. Archaeological evidence shows that *hospitium* associated with an Indo-Aryan institution practiced by Celtics and Germans among many other European ethnicities as well (Sanchez, 2001). The ancient hospitality was not pretty different from the modern one. In the days of the war, tribes wove alliances to move their armies altogether in case of external attacks whereas in peacetimes it encouraged the goods and travelers-exchanges (Korstanje, 2017). As Ramos and Loscertales (1948) puts it, hospitality was something else than a sign of inter-tribal friendship, rather it pointed to a much deeper process of solidarity that allowed the survival of smaller organizations before the stronger ones. Having said this, hospitality had two significant pillars which connected the politics with religiosity. The religious aspect of hospitality, Ramos and Loscertales adheres, unilaterally imposed the sacred mandate that strangers and pilgrims should be protected as absolute godsends. They represented a good opportunity to be in communion with Gods because, in the same way, strangers are treated, God will treat us in the hereafter. In ancient times, the popular parlance believed that natural disasters, famine, and other catastrophes resulted from the breach of the sacred law of hospitality. This suggests that the ontological security of human organizations depends on its ability to amalgamate their needs with God-will. The political side of hospitality, rather, was self-oriented to create a shield of protection against external attacks or invasions (Korstanje, 2011). Let's explain to readers that for the rise and expansion of modern nation-state, hospitality played a crucial role not only in conferring the right of free transit within and beyond Europe but also as the precondition for the emergence of mobilities and leisure travels (Pagden, 1995). Equally important, travelers experience greater levels of anxiety while traveling simply because they are visiting an unknown territory. At the same time, hosts are unfamiliar with the interests and aims of the newcomers. It is safe to say that hospitality helps to regulate the necessary risks in the host-and-guest encounters (Lynch, 2005; O'Gorman, 2006; Korstanje, 2011a, b; Andriotis and Agiomirgianakis, 2014; Korstanje and Tarlow, 2012). As Lynch et al. (2011) noted the current discussion in the academic circles is revolving around hospitality diverges in two contrasting poles. While some scholars define hospitality as a mechanism of surveillance oriented to strengthen the home security, others emphasize on its gift-exchange basis. Though the discussion is far from being closed, Jacques Derrida (2006) calls attention to the process of ethnogenesis as the platform in order for hospitality to surface. The sentiment of *us* is symbolically pitted against *them*, and this differentiation leads us to develop different strategies and tactics to accept or repel those who are not like-us. Although the presence of strangers allows nation-state to strengthen the in-group cohesion, no less true is that two types of hospitalities coexist: *Conditioned and absolute.* The latter demands the host to open the proper home to strangers without asking anything or expecting any gift in return. Rather, the former signals to that hospitality which is legally given only if guests give to me compensation. In this respect, it is important not to lose sight that Derrida clarifies, the subtype of conditioned hospitality rules not only in the modern world but also is enrooted in the nation-state, whereas the absolute hospitality remains as a utopia. Today's states unfold a process of selection to classify what travelers are accepted or rejected (Derrida and Duffourmantelle, 2000).

Last but not least, Daniel Innerarity (2001) alerts on the double-edge of hospitality for Western nation-states. Since risks should be equaled to hospitality, Innerarity adds, the quest for risk-zero societies inhibits our tolerance to uncertainty, as well as the respect for the undesired "Other." This accelerates a rapid moral decline that affects hospitality. The current protocols to find and eradicate risks impede to celebrate absolute hospitality (in terms of Derrida) in a way that our trust in the alterity not only declines but also paves the ways for the advance of uncertainty. The sense of hospitality is often practiced together a "human other" who questions us in a process of mutual reciprocity and trust. Here some pungent points arise. What would happen if the hosts are not humans? what type of hospitality can be granted by automats, machines, or androids? is this a token of the end of hospitality?

7.3   WHAT ABOUT THE SENSE OF REALITY AND FICTIONALITY?

The philosophical debate about the sense of reality and fiction seems not to be new. It can be very well traced back to Cartesian dualism as well as the works of medieval philosophers. For that reason, it is safe to say Descartes starts a new discussion that finally dissociates the mind from the body (Axtell, 2000). To put things simply, Descartes was originally motivated to show that the rationality of the mind and the emotionality of the body go for different paths. At the time the body equaled to a machine which works in automatic conditions, the mind controls the impulses and emotions that may lead us to our self-destruction. Governed by reason, and not by emotions, humans would reach a promising future. The concept of reality, for Descartes, was inexpugnable for the language, lest by the use of rationality (he compared to trial-and-error method (Descartes, 1991). Descartes ultimately introduces the theme of distortion as a key factor that motivates philosophers in their interrogations about life. Some pragmatists have questioned these assertions arguing that even carefully-planned research has obtained contradicting or false results, while unplanned or exploratory methods can obtain true findings (Musgrave, 1993). Richard (2000) offers an interesting model to understand this point. He says that the ontological subject elaborates its own image of the external world, not as a given object but as construction emanated from its inner-world. For him, the idea of reality is nothing more real than a concept to be recognized in the language. We are often immersed in stereotypes, irrationalities or prejudices that distort reality, as he convenes, but without the conceptualization of the object, the reality cannot be grasped. When this happens, the agency is placed asunder from the principle of reality. In consequence, the agency is immersed in a complex net of relations that give certain accessibility to reality. To put the same in other terms, we understand the world in parts, not as an all-catch concept. Over the years, constructivist philosophers alluded to the produced knowledge as an interplay of human construal while contradicting the ideal of objective reality as classic philosophy insisted (Jonassen, 1991). The rise of pragmatists who exerted a notable influence in the American philosophical tradition begins a new epistemological wave aimed at overlooking the influence of reality in the philosophical debates. For pragmatists, there is no important to ask whether reality exists or it does not, simply because the idea of reality was not marked in their agenda. To be more accurate, pragmatists emphasized on the belief that there is nothing real which cannot be named or apprehended by the language. We cannot understand those objects or things that our language does not name. It is noteworthy that the notion of reality not only was socially constructed but follows to psychological needs to have certainness in a chaotic world (Dewey, 1997). Therefore, it defies the classic concept of reality left by Descartes.

As the previous is the argument given, Richard Rorty (1982) clarified that pragmatism derived from an instrumental logic, which was historically forged in the ideological core of the American lifestyle. While for pragmatists the concept of truth was unilaterally subordinated to what is imagined or expressed in the language, the reality is manually molded according to what each person wants. This tradition, which is based on the economic instrumentality, fails to explain the nature of reality. W. Quine confronts with the pragmatism that any principle of verification comes from language. For Quine, empiricism not only much to say, but transcends the subjective meaning. Since Science and firstly Natural science seems to be the ultimate arbiter of "all knowledge," he polemically distinguishes science from philosophy. While the former is a single system which subordinates to all disciplines under the logic of physic or the natural world, the latter rests on the monopoly of beliefs. His position interrogates on a new theory of knowledge and reality, which was continued by different voices as Donald Davidson.

In his book, *Truth, Language, and History*, Donald Davidson (2005) alludes to the limitations of pragmatism, which not only neglected the concept of reality but also introduced a biased diagnosis of language. Most particularly, though objects can be ingrained in the language, through the articulation of meaning, their existence cannot be denied. The paradox lies in the same conception of language pragmatists and deflationists employed. To a closer look, the language helps to express some intention from the speaker to the hearer. This is a process of symbolism, not a creator of reality. For the sake of clarity, if I say the moon is white, we understand that such a sentence is simply true because the moon is white. This, Davidson adds, is a tautology of the language which should be overcome. Davidson's main thesis is that the concept of truth is something else than the properties of their constituents.

> *"We are interested in the concept of truth only because there are actual objects and states of the world to which to apply it: utterances, states of belief, inscriptions. If we did not understand what it was for such an entity to be true, we wouldn't be able to characterize the contents of these states, objects, and events. So, in addition to the formal theory of truth, we must indicate how truth is to be predicated of these empirical phenomena" (Davidson, 2005, p. 35).*

In a thorough review of Davidson and Quine; Glock (2003), from the University of Reading UK, said that there is a substantial difference between both thinkers. The former alludes to reality as the interplay of mental events, which are culturally internalized through the use of language, while the latter signals to the natural world as the ultimate border between selfhood and its world. Whatever the case may be, the dissociation between emotions and rationality, in the same way, that truth and fictions were historically centered to legitimate a paternalistic viewpoint, which was certainly forged during the colonial rule, to deepen the center-periphery dependency. While Europeans were considered the peak of culture and civilization, rational, educated, and illustrated, the aborigines were historically considered backward, illiterate, emotional or emotionally thoughtless (Korstanje, 2012). The West fleshed out a concept of progress not only in the European rationality but in the monopoly of technology which was historically functional to colonialism. In the same way, rationality controls emotionality, white Lords should educate wild savages. The idea of progress was the touchstone of a civilization that adventured to discover and control the external periphery. In so doing, emotionality was limited to be only enrooted in the culture of entertainment. Interesting studies have revealed how literature was manipulated to exploit the nonwestern others as an object of gratification in times of colonialism (MacCannell, 1984, 2011; Jeffreys, 1999; Bandyopadhyay and Morais, 2005).

Lastly, the notion of fiction and objectivity was brilliantly questioned by professors Lamarque and Haugom Olsen (2002) in their book *Truth fiction and literature. These specialists* proffer a radical reading on the classic definitions of reality left by philosophy and the romantic viewpoints of literacy. The nature of literary value, as linked to the humanistic idealization of life, fabricates cultural products (as stories or novels) that shed the light on specific situations. Like movies, the literary grounds interrogate us furtherly some issues otherwise cannot be understood. For that reason, the understanding of movies' plots helps us to study our society from the inside. Is the use of robots emulating the same colonial ideologies or discrepancies between rationality and bodies' pleasures?

7.4 IN ROBOTS, MUST WE TRUST?

Recently, some specialists explored the role of technology as the main driver towards innovation in tourism and hospitality fields. The success or decline of a tourist destination depends on the flexibility to adapt the market demand, which includes customization, technological innovation, and entrepreneurship (Buhalis, 1998; Yuan, Gretzel, and Fesenmaier, 2006; Buhalis and Law, 2008; Hjalager, 2010). Because of time and space, we shall leave the topic of innovation for another moment. Instead, we shall focus on the role of robots as configurators of a new type of hospitality in a futurist world. The first point of entry in this discussion relates to the pleasure of maximization and hedonism, which seem to be one of the main goals pursued by modern holidaymakers. In this token, Yeoman et al. (2012) call the attention that tourism has situated as a force of global growth while it expected to continue to grow in the decades to come. For some reason, tourism, and technology were inextricably intertwined. Furthermore, science fiction should be understood as a set of speculations and narratives that portray some realistic assumptions and new epistemological borders of our daily existence. The adoption of ICT and robots are not only revolutionizing the tourism and leisure industries but also speak of the interconnection between the selfhood and the alterity. We are having some ethical problems to legalize some trouble practices, above all around sexuality such as prostitution and child abuse. Sex robots in the tourism industry would lead us to avoid serious ethical discussion liberating some vulnerable actors for a cruel system of exploitation.

As the previous backdrop, Yeoman and Mars (2012) emphasize that technology has become in an integral aspect of the human condition. It is a clear mistake to deepen the division between humans and androids. Doubtless sex and the quest of perfect sexual arousals are not exemptions. Beyond the countless solutions, sex robots offer to humans as the VIH transmission, human trafficking, and prostitution some ethical issues should be placed under the critical lens of scrutiny. Why should people pay for sexual services?

Though the reasons behind the impulse of persons to have sex with robots are not clear in Yeoman's texts, the futurist scenario offers a dissociation between ethics and practices. This begs a more than an interesting philosophical point, to what extent the use of robots for (erotic) pleasure-maximization is ethical? As Jennifer Germann Molz (2014) puts it, technology may be very well used or misused by tourists following the mandate of ethics, the fact is that digital technologies have caused a fast acceleration not only in the transport timeframe but also in the landscape and how the postmodern travels are performed.

In the post COVID-19 context, some theorists emphasize on the importance of robots to enhance a secured social distancing between hosts and their guests. In a recently-published note of research, Seyitoglu and Ivanov (2020) enumerate the potential benefits of introducing robots in psychical distancing for the tourism industry. Per these analysts, though robots help to potentiate security in the direct contact, controlling the virus dissemination, no less true is that "robots" recreate a wall between tourists and employees which may affect the emotional closeness.

The introduction of digital technology and our enthusiasm for media have altered the ways the other is perceived, the surrounding sensible world as well as the rise of new more genuine forms of hospitality which were unknown a couple of decades back. In fact, she cites the case of couch-surfing where tourists are freely lodged strengthening the necessary reciprocity to be welcomed by others at a later day. Although some critical voices (Marxists and Post-Marxists) bemoaned that the use of technology in tourism would be causing an alienator effect on tourists the evidence suggests more harmonized forms of hospitality and relations are emerging. What is more than important to discuss is the connection of digital technology with the creation of virtual reality or pseudo-landscapes affordable for tourists at the comfort of home. Traveling without moving and gazing without traveling are two of the main trends future tourism is demanding.

The question of ethics is not a minor issue. As discussed in other chapters, Jost Krippendorf envisaged tourism as a mechanism of escapement which served not only for daily frustrations in lay-citizens to be revitalized but also to keep society working. Unlike other colleagues, he held the thesis that tourism is not good or bad, it follows the cultural background of society (Krippendorff, 1987). The issue is very well illustrated by Jim Butcher who holds that nobody can be against ethical tourism, unless by the fact sometimes this is symbolically associated with small enterprises instead of producing holistic changes. The ethical problems should be tackled, changing the mainstream cultural values that generate them. Still further, the popular parlance accepts ethical tourism in small-scale while recognizing that the opposite which ranges contamination, human rights violations, or unethical practices, can be reserved to mass-tourism. In consequence, ethics have been commoditized as the best course of action for ONG as a strategy of marketing but ignoring the voices of local in the process. Last but not least, most of the behavior or performance which are considered as unethical for specialists such as acculturation process or cultural shock are not crimes but only the necessary consequence of host-guest encounter. This confusion happens because these specialists think in cultures as static boxes that remain inexpugnable to human grasp. Butcher toys with the belief that cultures are in constant change anytime and everywhere. The rise of tourism and mobilities has activated constructive synergies that resulted in positive effects with the local communities (Butcher, 2009).

7.5   FUTURE TOURISM: THE PASSENGERS

For summing up the philosophical core of this discussion, we need to review the film plot *The Passengers* which shows the cruel intersection between robots and what we dubbed *"a failed hospitality."* *The Passengers* is released through the end of 2016 and was originally starred by Jennifer Lawrence, Chris Patt, Michael Sheen, and Laurence Fishburne. This should be cataloged as a science section film. The plot situates in a futurist landscape where the starship *Avalon* should transport 5.000 colonists and 258 crew members to the planet *Homestead II.* Because of the distances and traveling times, the passengers are placed in hibernation pods for almost 120 years. One of the crew members the mechanical Jim Preston is awakened 90 years after the departure in view of a serious malfunctioning. He (Preston) wanders through the starship without human companion unless by an android bar-tender which bolsters a fluid communication with him. Arthur and Preston became in good friends; but though Preston tries, he feels isolated and alone. This happens until he notices on the presence of a youth writer, Aurora Lane (Lawrence) in the same ship. Above all, Arthur is not human and though he was programed to serve Preston in everything he needs, he has no soul, no will.

In what follows, Preston scrambles with his own morality whereas he makes the decision to awaken Lane from her hibernation pod which ensures life in common or suicide for both. Arthur knows that he is the only one living human in this ship, and of course, he needs a human companion. After further mediation, Lane is revived because Preston fell in love with her. She thinks that everything was a result of the same malfunctioning brought Preston from the slumber he was. Inadvertently, Arthur said the true to Aurora after falling in language misinterpretation. Needless to say, she is angry while Jim tries to apologize. In the meantime, a third person, Gus (Fishburn) is awakened up by a failure in the software of the ship. Such a malfunction was a consequence of a failure in the fusion reaction unless rapidly fixed will end with the life of all passengers. Ultimately, all this mess has a sense, and Jim immolates attempting to fix the reactor from the exterior in order for his loved Aurora to be salved. What started with an act of egoism ends with the most sublime proof of love. Jim is dead but he resurrects him using the biotechnological instrument of the starship. Aurora can now return to the pod to proceed the trip but Jim has not the same luck. She makes a difficult decision and embraces a happy life with Jim though she never arrives at Homestead II. Around 88 years later, the crew members awake to find a small house in vegetation in the concourse area. Just their Aurora's voice indicates her decision to be jointly to Jim to know what love is. She left a book that narrates her life with Jim. What does this film teach us or what is the lesson to be learnt after describing *The Passengers*?

The external deep space, as well as the university, remains as one of the greatest mysteries of mankind. Traversing the borders of the earth in space flights is today "science fiction." However, it emulates the same vulnerabilities of tourists when they should move towards an uncertain destination. What is equally important, hibernation evinces the epistemological borders of reality and fictions or the wakefulness and sleep. The real happiness is not determined by the quest of selfish pleasures conditioned hospitality may give (for example when Jim awakens Aurora) but in the freedom only reality ensures. Besides, the support of high tech is vital to use robots as support staff in these types of travels, but hospitality needs from a "human other" to engage. The fact is that this is the nature of hospitality. The passengers, as well as the introduction of Robots in the tourism industry, are a clear sign of an anti-hospitality sentiment generated just after 2001. Likewise, Jim was empty without the companion of Aurora, though Arthur was disposed to fulfill all their needs, hospitality activates a sentiment of reciprocity with the Other. Jim has in the starship everything that he dreamed, and Arthur offered exquisite liquors, and finest whiskies but Jim needed something else. The thought struck him that most likely Aurora should be his companion. It followed a selfish behavior but with the passing of time, Aurora feels the same for Jim conforming in the shoot an adorable-looking pair. That is important to add that this love rests on a terrible lie that Arthur unwittingly discovers. This is precisely the logic of robots, they do not lie, they do not cheat, they do not decide beyond what they have been programed, after all, they lack of autonomy to decide. This represents one of the moral dangers to instrumentalize robots to meet humans' pleasures.

The *Passengers* reminds not only the importance of alterity to conform to the selfhood but also hospitality is never feasible without human presence. While technology always can fail, human autonomy to decide what is wrong or good remains. Although robots can be of importance and helpful in future travels, hospitality seems to be what defines us as humans.

Another important point of entry in this discussion is given by HBO series, Westworld where guests are invited by a great corporation to maximize their pleasure no matter the ethical consequences. Tourists travel to Westworld to gain novel and fascinating experiences. However, hosts are not real humans, they are humanoids perfectly designed to meet all the desires of their guests. In perspective, hosts are fabricated robots subordinated to cruel exploitation in the staged-landscape that emulates the tough life in the far West, finally struggle for their freedom due to some problems in their configuration. Westworld not only exhibits the logic of modern capitalism but also the end of hospitality as a platform where hosts and guests not only interact but also each one understands the "Other." For the genuine hospitality prevails, the host-guest meeting should be based on egalitarian conditions.

7.6　CONCLUSION

From its onset, the Western social imaginary has problems to understand the "Non-Western Other". There was a dissonance between rationality and emotionality which acted discursively to dominate the natives. The idea of the noble savage spoke us of the urgency of expanding Western civilization to the world. Because of COVID19, the figure of the "Other" has been demonized and neglected. With the industry on the brink of the collapse, scholars interrogate on the future of tourism as well as the ethical limitations of introducing robots to mediate between hosts and guests. While the first chapters debated the nature and evolution of tourism industry and its connection with modernity, this chapter focuses on the role of the morbid gaze, and the end of hospitality. We have discussed the ethical dilemmas of disposing of robots or humanoids to maximize guests` pleasure. With focus on the plot of the Passengers, the chapter holds that digital technologies engender standardized and mechanical forms of solidarity that affect seriously the Sacred Laws of Hospitality. All these topics will be addressed in the next and final chapter.

KEYWORDS

- anthropology
- artificial intelligence
- COVID-19
- hospitality
- mobilities
- otherness
- tourism industry

REFERENCES

Andriotis, K., & Agiomirgianakis, G., (2014). Market escape through exchange: Home swap as a form of non-commercial hospitality. *Current Issues in Tourism, 17*(7), 576–591.

Axtell, G., (2000). Introduction. In: Axtell, G., (ed.), *Knowledge, Belief and Character: Reading in Virtue Epistemology* (pp. xi–xxix). Rowman & Littlefield Publishers Boston, MA.

Bandyopadhyay, R., & Morais, D., (2005). Representative dissonance: India's self and western image. *Annals of Tourism Research, 32*(4), 1006–1021.

Buhalis, D., & Law, R., (2008). Progress in information technology and tourism management: 20 years on and 10 years after the Internet—the state of e-tourism research. *Tourism Management, 29*(4), 609–623.

Buhalis, D., (1998). Strategic use of information technologies in the tourism industry. *Tourism Management, 19*(5), 409–421.

Butcher, J., (2009). Against 'ethical tourism'. In: Tribe, J., (ed.), *Philosophical Issues in Tourism* (pp. 244–253). Bristol Channel View Publications.

Davidson, D., (2005). *Truth, Language, and History.* Oxford University Press, Oxford

Deckard, S. G., (2007). *Exploited Edens: Paradise Discourse in Colonial and Postcolonial Literature.* Doctoral dissertation, University of Warwick.

Derrida, J., & Dufourmantelle, A., (2000). Of hospitality: Cultural memory in the present. *Trans. Rachel. Bowlby.* Stanford University Press, Stanford, CA.

Descartes, R., (1991). *The Philosophical Writings of Descartes: Volume 3, the Correspondence.* Cambridge University Press, Cambridge.

Dewey, J., (1997). *How we Think.* Courier Corporation, New York, NY.

Gale, T., (2009). Urban beaches, virtual worlds and 'the end of tourism'. *Mobilities, 4*(1), 119–138.

Germann, M. J., (2012). *Travel Connections: Tourism, Technology and Togetherness in a Mobile World.* Routledge, Abingdon.

Glock, H. J., (2003). *On Language: Thought and Reality.* Cambridge University Press, Cambridge, MA.

Hjalager, A. M., (2010). A review of innovation research in tourism. *Tourism Management, 31*, 1–12.

Innerarity, D., (2001). *Ética de la hospitalidad. [Ethics of Hospitality]*. Península, Madrid.

Jeffreys, S., (1999). Globalizing sexual exploitation: Sex tourism and the traffic in women. *Leisure Studies, 18*(3), 179–196.

Jonassen, D. H., (1991). Objectivism versus constructivism, Do we need a new philosophical paradigm? *Educational Technology Research and Development, 39*(3), 5–14.

Korstanje, M. E., & Tarlow, P., (2012). Being lost: Tourism, risk and vulnerability in the post-'9/11'entertainment industry. *Journal of Tourism and Cultural Change, 10*(1), 22–33.

Korstanje, M. E., (2011b). The fear of traveling: A new perspective for tourism and hospitality. *Anatolia, 22*(2), 222–233.

Korstanje, M., (2011a). Reciprocity, hospitality and tourism: An examination of Marshal Sahlinss contributions. *European Journal of Tourism, Hospitality and Recreation, 2*(2), 89–103.

Korstanje, M., (2012). Reconsidering cultural tourism: An anthropologist's perspective. *Journal of Heritage Tourism, 7*(2), 179–184.

Korstanje, M., (2017). *Terrorism, Tourism and the End of Hospitality in the West.* Palgrave Macmillan, Basingstoke.

Krippendorf, J., (1987). *Holiday Makers.* Butterworth Heinemann, Oxford.

Lamarque, P., & Haugom, O. S., (2002). *Truth, Fiction and Literature.* Oxford University Press, Oxford.

Lynch, P. A., (2005). The commercial home enterprise and host: A United Kingdom perspective. *International Journal of Hospitality Management, 24*(4), 533–553.

Lynch, P., Molz, J. G., Mcintosh, A., Lugosi, P., & Lashley, C., (2011). Theorizing hospitality. *Hospitality & Society, 1*(1), 3–24.

MacCannell, D., (1984). Reconstructed ethnicity tourism and cultural identity in third world communities. *Annals of Tourism Research, 11*(3), 375–391.

MacCannell, D., (1992). *Empty Meeting Grounds: The Tourist Papers.* Routledge, New York: NY.

Merton, R. K., (1968). *Social Theory and Social Structure.* Simon and Schuster, New York, NY.

Musgrave, A., (1993). *Common Sense, Science and Scepticism: A Historical Introduction to the Theory of Knowledge.* Cambridge University Press, Cambridge.

O'Gorman, K. D., (2006). Jacques Derrida's philosophy of hospitality. *Hospitality Review, 8*(4), 50–57.

Pagden, A., (1995). *Lords of all the World: Ideologies of Empire in Spain, Britain, and France c. 1500-c. 1800.* Yale University Press, New Haven, CT.

Paul, R., (2000). Critical thinking, moral integrity and citizenship. In: Axtell, G., (ed.), *Knowledge, Belief and Character: Reading in Virtue Epistemology* (pp. 163–176). Rowman & Littlefield Publishers, Boston.

Quine, W. V. O., (2013). *Word and Object.* MIT Press, Cambridge, MA.

Ramos, Y. L. J., (1948). Hospicio y clientela en españa cética. *Revista Emérita,* [Hospitality and Tribes in Ancient Spain]. *10*(1), 308–337.

Rorty, R., (1982). *Consequences of Pragmatism: Essays, 1972–1980.* University of Minnesota Press, Minneapolis, MN.

Sánchez-Moreno, E., (2001). Cross-cultural links in ancient Iberia: Socio-economic anatomy of hospitality. *Oxford Journal of Archaeology, 20*(4), 391–414.

Seyitoglu, F., & Ivanov, S., (2020). Service robots as a tool for physical distancing in tourism. *Current Issues in Tourism.* doi.org/10.1080/13683500.2020.1774518.

Yeoman, I., & Mars, M., (2012). Robots, men and sex tourism. *Futures, 44*, 365–371.

Yeoman, I., Tan, L. Y. R., Mars, M., & Wouters, M., (2012). 2050 *Tomorrow's Tourism.* Channel View Publications, Bristol.

Yuan, Y. L., Gretzel, U., & Fesenmaier, D. R., (2006). The role of information technology use in American convention and visitors bureaus. *Tourism Management, 27*(2), 326–341.

CHAPTER 8

The Impact of COVID-19 in the Tourism Industry: End or Rebirth?

ABSTRACT

Is tourism in crisis or this crisis is evincing the end of tourism? After several alerts given by the outbreak of different viruses such as H1N1, SARS or MERS and the potential danger they represent for the industry, scholars overtly said that tourism exhibited a serious risk for the dissemination of a new global pandemic. This was particularly true with the COVID-19 outbreak in Wuhan-China. In months the COVID-19 was not only disseminated through Europe and the US but paralyzed the global commerce even the tourism industry worldwide. Tourism grids into a halt a paralyzed crisis there are no records since the Spanish flu. This chapter deals with the problem of emergencies and crises and how they can be utilized as new opportunities for the rebirth of the industry. The different sections of this piece synthesize the recent advances in emergency and disaster studies. In a world without tourists, tourism research shows fewer opportunities to survive.

8.1 INTRODUCTION

Coronavirus (or COVID-19) is an infectious disease caused by a virus (SARS-CoV-2) found initially been in Wuhan, China, and globally spread. The common symptoms include fever, cough, and shortness of breath. To date, on June 08, the virus has infected 35.452.877 person, causing globally 1.042.732 victims. The most affected countries are mainly the US (214.629 deaths), followed by Brazil (146.732), India (102.476), Mexico (79.088), the UK (42.350), Italy (35.986), Spain (32.086) and France (32.230), but these ciphers are in constant evolution. COVID-19 not only shocked the world because of its rapid dissemination but also pressed to different governments to adopt some radical measure as the closure of borders, the suspension of all travels and a radical quarantine for lay-citizens. Tourists who were traveling abroad were unilaterally repatriated or jailed if they reject to be returned to its home. These measures were accompanied with new ones such as the closure of borders and airspace or the cancellation of commercial flights. To put simply in other terms, COVID-19 not only affected tourism as other similar viruses such as SARS, or Swine Flu but also was the last straw for the tourism industry which suffered millionaire losses. The first doctors, who originally discovered the disease, were censored by the Chinese government through December 19. Travels coming from China arrived in these months to Northern Italy to be the epicenter that infected all Europe. The decision made by European authorities to close off all affected airspace pressed the US to do the same. Those tourists who do not respect the quarantine in the US or Latin America were fined or expatriated. Doubtless tourism and hospitality industries enter a difficult crisis as never before. A book about tourism needs a final chapter dedicated to COVID-19 though at this stage it is still too early to reach substantiated conclusions.

In the first section, we review the nature of the emergency in the lens of sociological insight. While policy-makers design carefully crisis management protocols, it is very difficult to follow them because it is the nature of a crisis. Paradoxically, the protocols fail at the time the normal conditions of life are suspended. The second section is a summary snapshot of some authoritative voices who have alerted on the needs of adopting crisis management campaigns in the fields of tourism and hospitality. Last but not least, we introduce readers to a critical debate revolving around the future of tourism as well as the challenges the industry faces after COVID-19 (Coronavirus disease).

8.2   THE NATURE OF THE CRISIS AND THE STATE OF EMERGENCY

One of the fathers of disaster studies, Enrico Quarantelli argued convincingly that protocols always fail in case of disasters or in states of emergencies. He starts from an interesting point: why do experts fail to forecast disasters or made bad decisions to mitigate their effects?

As Quarantelli notes, experts or policy-makers, who have been educated to act in emergency cases, make the incorrect decision, not by lack of familiarity with the issue or inexperience, but precisely because of the nature of the disaster. Per his viewpoint, a disaster or a crisis seems to be an unexpected event that disrupts all necessary protocols and the responsiveness of society (Quarantelli, 1985, 2005). In the state of emergency, the constitutional rights are commonly suspended while the executive branch received further delegation, authority, and power than in normal life. This moot point was widely studied by sociologists who alert on the problem it has for the democratic system (Scheueman, 1999; Sunstein, 2002, 2005; Honig, 2009). The state of emergency should be temporally regulated in order for the democratic institutions not to succumb to a despotic leader. Michelle Foucault (2003) explains that society keeps altogether through an invisible hand, *the power of discipline*. Here, the term discipline does not equate to repression, but to the capacity to control and accept emerging risks. He remarks that risks are similar to vaccines, which are inoculated viruses. The vaccine does not eradicate viruses; rather it dispossesses them from the lethal characteristics. For Foucault, society orchestrates countless disciplinary mechanisms to regulate external threats while they are incorporated in a nuanced version. However, may this sentiment of fear being invented or politically manipulated?

In his book, *Governing through Crime*, Jonathan Simon (2007) critically reviews the conception of emergency and criminality which were historically manipulated in order to provide to the American Executive branch with further legitimacy. When this happens, politicians incur in terms such as *the struggle, the war, or campaign.* The senate confers further faculties to the president to intervene in democratic institutions at its discretion. The unlimited transference of power to the president would be particularly risky for the democratic life. This has been observed in the 70s decade, when the president declared the war against cancer, a carried on in the decades to come such as the war against crime in the 80s, or today with the war against terrorism (Simon, 2007). It is safe to say that we can affirm that the world is in the war against COVID-19. Hoskins and O'Loughlin employ the term CNN effects to mention the climate of fear often the media recreates to cover disasters or disrupting events. Today's media not only processes information but also covers distant event otherwise cannot be reported in other epochs. Such a mediatization of the world is certainly based on the digital technology disposed to cover a distant event which makes closer to a global audience. Jean Baudrillard coins the term *The Spectacle of terror* to denote the constant and durable state of emergency created to govern the present-from the future. He cites the plot of the movie *Minority report* to explain how capitalism has created a hyper-reality which never takes place in the real world. The mediatized events do not exist in reality, as Baudrillard adds. Minority report speaks us of a utopian future where Police successfully reduced crime to zero. With the *Precogs (mutants who have the providence to see the future)*, Police anticipate to crimes earlier they are really committed. As Baudrillard puts it, the same applies to the risk culture which governs the present through the invention of risks never happen. The risk, at least in its nature, operates in the future but with palpable effects in the present.

As the previous argument is given, Paul Virilio, in different essays, calls the attention on the interlink among terror, media, and technology. Virilio concentrates efforts in deciphering the role of image in the formation of modernity and how it affects daily life. He sustains that mass-media exerts a notable influence over global audiences, a power which is inexpugnable for the politicians. While humans have a natural disposition to communicate with others, no less true is that media fabricates, packages, and disseminates far events otherwise would remain unknown. This geographical dispersion is replaced in a flat-screen 24 hours per day. The communication disposes people in egalitarian condition, it closes them but when the communication is processed by the media, this closeness is effaced. As a consequence, people live physically closer while psychologically farther. The digital technologies and media distance citizens from their institutions. Above all, the recent technological breakthroughs have accelerated the transport, travels, and leisure time, creating a gap which is filled by the media. The excess of velocity implies a sentiment of inferiority and fear where the self believes the future can be controlled and dominated. In this respect, Virilio toys with the belief that faster people travel more they believe they are controlling the future. The quest for something new, a sentiment very characteristic of excited tourists, really distances from self-introspection. The classic anthropology operated in the notion of here-and-we and there-and-they but these concepts do not apply anymore in the modernity. The wall protected the city in ancient times, but now the cities are being inter-connected without frontiers. The notion of there has inverted to here. The accident is commoditized in the form of spectacle to control not only the body but also the desire. We consume disasters to terrorize us and in so doing we cannot stop to do it. The market does not promote the reason, but the emotions oriented to consume disasters. Having said this, fear is certainly based in the excess of information which and the imagination emptiness. The causality of probabilities-beyond the events-is replaced by a deeper logic where the panic monopolizes the social relations. The Science which historically served to human protection is adapted to be at the disposal of the market. Laboratories and universities do not work to prevent risks but to mitigate their economic effects in the market (Virilio, 1995, 2005, 2010). Undoubtedly, the tyranny of speed has changed the geography forever, mediating among consumers.

Anthropologists traveled to overseas colonies to be there with the non-Western natives. Nowadays, natives have migrated to the urban cities forming a new emerging consuming class. The relation self-alterity has set the pace to a radicalized emptiness without borders and time. Marc Augé recognizes that the anthropology, which was a discipline thought to understand the other, is changing its own foundations in an ever-mobile world where the *anthropological place* subverts to the *non-place*. The question of whether the place is defined as a relational space of reciprocity and identity, the *non-place* refers to a space of consumption and depersonalization lacked from identity and tradition. Examples of non-places are everywhere, airports, highways, train stations, shopping malls (only to name a few). Tourist's identity is only checked out when its passport should be presented. This led to thinking Auge that a non-place is saturated of the present but closed to the pastime. There is an excess of ego which inhibits social relations. It is important not to lose the sight that the panic would be a reaction emanated from the abundance of anonymity. The dissociation between past and present re-symbolizes the human networks while (under) modernity alters the geographical distances. The "Other" appears not to be now a foreigner unknown, frequently frightened because he or she looks very different. Now this "Other" is feared because of its closeness to us. The urban areas devote considerable efforts to attract foreign tourists experiencing a saturation of images and consumerism that invariably leads towards a process of depersonalization. The crisis of sense that Occident experiences by these days are no other thing than a crisis of alterity, as Augé reminds. Ethnographers have difficulties in acknowledging that the alterity overrides the alterity. By the articulation of stereotyped narratives, these tourists are in quest of a fictionalized world which only can be grasped (if not understood) through consumption. People are not traveling to discover; they are moving to consume an externally-designed cultural product. He coins the term *The impossible trip* to the tourism industry. Tour operators not only re-draw the world but also mold the sensations behind the experience. Likewise, the (imagined) travel is sold before to be made. Tourist travel is a travel that never starts, he adheres. For Augé tourism should be understood as a fictionalized travel that shows the inequalities of society. Privacy is disposed to the needs of controlling the conflict. Whether classes are segregated in daily life, for example, elder people are confined to geriatric while rich citizens dwell fenced-off neighborhoods, at the beach all, are, and do the same. The tension between the unknown and the known that characterized the anthropological travel is superseded by tourist travel. Like the tourist travel has no connection between the departure and the arrival, the media blurs the relation between causes and consequences rising anxiety and fear in society (which explains the current state of crisis) (Augé, 1995, 1998). Far from being a foundational event, COVID-19 marks the end of a process, the radical mutation of tourism and hospitality to a hybridized form. For some reason, the economic-based theory occupied a central position in the configuration of tourism episte-mology. The paradigm not only punctuated on the needs of controlling those variables which affect the attractiveness of tourist destinations but mistakenly speculated on the voice of tourists as the only valid source of information. The urgency of measuring was imposed to the needs of understanding, as John Tribe (2010) laments. However, the large volume of publications, associated to the interests of social scientists for tourism as object of study was not enough for the discipline to be considered a serious option. In addition, quantitative methodologies gradually enthralled as the only valid instrument to understand the tourist's desire. As Gale (2007) eloquently puts it, it is unfortunate that the history and tradition of tourism research is mainly centered on the biases and prejudices of positivism which was accompanied by a managerial viewpoint of the discipline. In a complex and ever-changing world quantitative method do not suffice to explain the evolution of tourism, at least in a coherent way. What is more important, in a world without tourism, what are the opportunities of the economic-based paradigm to survive? is COVID-19 the clear sign of the end of tourism research?

8.3   TOURISM AND CRISIS MANAGEMENT

The literature that deals with crisis management can be divided into two clear-cut poles. Those who center on the problem of terrorism, a sub-discipline born through 90s decade (Somnez, Apostolopoulos, and Tarlow, 1999; Stafford, Yu, and Armoo, 2002; Blake and Sinclair, 2003) and affirmed after 9/11 and those studies that study the reaction of tourism to different natural or made-man disasters such as earthquakes, hurricanes, or virus outbreaks (Laws, Prideaux, and Chon, 2007; Ritchie, 2004; Santana, 2004; Anderson, 2006). Although both traditions focus on crisis management as the main commonality, they have serious differences. While the former signals to the urgency to prevent the future risks, the latter emphasizes on the adaptability of the system to mitigate the disastrous effects of the event. In this vein, Somnez, Apostolopoulos, and Tarlow (1999) suggest that terrorism-to some extent-is more harmful than natural disasters, simply because when persistent affects negatively the destination's image declining its attractiveness for a long time. Authors go on to write:

"Fewer terrorist incidents in the United States have been recorded for the first half of the 1990s; however, their nature and magnitude are more severe than those of past years" events. Experts indicate that terrorism will continue to victimize "soft" targets, attacks will become more indiscriminate, terrorism will become institutionalized as a method of armed conflict, it will spread geographically, and the public will witness more terrorism than ever due to the medias improved ability to cover terrorist incidents"
                    (Tarlow, Apostolopoulos, and Tarlow, 1999, p. 14).

For Professors Adam Blake and Thea Sinclair, tourism crisis management should be adopted in moments when the effects are durable and extendable in the time. This was particularly the case of the US and the tourism downturn in the post 9/11 context. The repercussions of terrorism are even extended to other service sectors than tourism such as airplanes, hotels, or restaurants. Most probably, the government invests considerable resources to placate the crisis. The duration of the downturn, as well as the decision revolving around how important or feasible the intervention is two of the steps to follow in cases of emergency. Tourism safety and security is generally sensitive to health concern but the main problem with terrorism lies in its political nature and the instilled fear which may put a destination in a symbolic quarantine for months or years.

In fact, beyond the effects of a catastrophe in a city or town, terrorism has the capacity to create long-lasting consequences through the dissemination of fear. The same applies to the crisis management adopted by countries that suffered a lethal virus outbreak. Blake, A., & Sinclair, M. T., (2003) studied the sudden and unsettling outbreaks of SARS (severe acute respiratory syndrome) to explain in what degree the tourism industry was harmed in Singapore. They hold that the hotel sector experienced major material losses while proposing helpful guidelines to adopt tourism crisis management. Echoing this, Dirk Glaesser (2006) argues that tourism faces one of the most exciting challenges in the turn of the century. Though global virus outbreak was certainly uncanny in the past, the theme recurs frequently in the present. The higher levels of mobilities in the first world associated with the technological advances, associated with the growth of tourism make the adoption of crisis management a priority. To wit, Brent Ritchie (2004) alludes to chaos and disaster as important factors to achieve a more resilient tourism industry. For some reason, scholars systematically overlook the figure of chaos as unpredictable and a variable very hard to grasp, but indeed, we cannot understand the nature of disaster without approaching chaos and crisis. He eloquently enumerates three types of crises, *immediate, emerging, and sustained crises.* The immediate crisis corresponds with a preliminary stage where organizations have no necessary warning to react. The emerging crisis consists in the gradual development of the crisis which can be stopped. Finally, the sustained crisis speaks to us of durable effects on society and tourism.

As this backdrop, Mariana Sigala (2011) turns her attention to the communicational campaign to contain the fear, doubts, and worries these types of events stir up. She exerts a radical criticism to the specializing literature (with a focus in crisis management) showing that under some conditions the excess of information accelerates the crisis. Policy-makers should analyze the steps or measures to follow according to the type of event, the phase of an emergency, the number and nature of stakeholders involved and the geographical area in question. Speakman and Sharpley (2012) get interesting conclusions in the case of Mexico and the H1N1 virus (Swine Flu). Like Ritchie, Speakman, and Shapley invite to consider chaos as the main variable in crisis management research. Since tourists are typically averse to risk, it is important not to lose the sight of the fact that any failure in the health system to contain a virus has practical aftermaths to the time of avoiding a certain infected destination. Virus outbreaks often spread rapidly canceling flights while blocking tourist fluxes. The possibilities of the specialized literature to give solutions and answers to policy-makers remain at least questionable.

"The purpose of crisis and disaster management models is, evidently, to provide guidance to destination and business managers and planners prior to, during, and after a crisis event. In specific circumstances, this objective has been achieved. However, the extent to which these models more generally represent realistic, practical responses to crisis situations is limited by a number of factors"   *(Speakman and Shapley, 2012, p. 2).*

Unlike palpable risks, viruses are microorganism which cannot be seen, grasped waking up the higher sentiment of uncertainty and fear making that the destination's attractiveness declines and the tourism industry collapses in days. Quite aside from the role played by the communicational factor or the chaos, the reviewed studies are based on the economic factor only misjudging other relevant aspects of the issue. COVID-19 not only harmed tourism but prompted an inevitable downturn since the stock and market crisis in 2008. This pandemic, which can be considered a foundational event comparable to the Second World War or 9/11, shifted critically the normal fluency and mobilities of the first world. Hotels and restaurants are today being refashioned to be used as hospitals to deal with infected people while commercial airplanes are being disposed to repatriate co-national tourists stranded abroad. In the next section, we shall analyze what other scholars dubbed as *the end of tourism.*

8.4   COVID-19 AND THE END OF THE TOURISM INDUSTRY

In their seminal book, Economies of Signs and Spaces, Scott Lash and John Urry (1993) alert that the classic Fordism or mass-scale production sets the pace to a new global mean of production which is decentralized and virtual. Above all, we do not consume products, we are consuming sings. When we drink a coffee from Colombia, or a whisky from Scotland, the sign molds the experience. This happens because late-capitalism has re-organized the means of production according to a sign that mediates between commodities and consumers. As a social act, tourism not only is dependent of the geographical displacement but engages with a temporal rupture with the routine. In Lash and Urry, tourism would be never possible at other times than modernity. The tourist gaze expropriates the "Others" from their constitutional characteristics creating signs and ways to interpret them. The trajectory (exchange) of goods and consumers not only leads people into a fictional landscape but accelerates the inevitable decline of solidarity and reciprocity. The question of whether of Maussian theory , which focuses on giving-while-receiving circuits, offers an interesting model to understand the economy and social solidarity, one might speculate that trade affects directly the human reciprocity. The tour operators are trained to give security to their clients. They know the destination they are offering while absorbing the potential risks tourists would face in their vacations. The net of experts plays a leading role in codifying and de-codifying those symbols that ultimately ensure the tourist's experience. In the 19th century, Occident went through a reformulation of a new romanticism opening the doors for the rise of flexibility process where the "Other" situated as an object of desire and consumption. The economies of signs evolved together the economies of desire stimulated and regulated by modern tourism. For Lash and Urry, the end of tourism is only a question of time, and of course the result of a decentralized economy which is globally inserted in the power of signs. Echoing this claim, Tim Gale (2008, 2009) asks furtherly on the end of tourism as the interplay between the inflation of risk perception and the emergence of virtual reality. The hegemony of virtual reality and the current technological breakthroughs adjoined to the anti-tourist's movement, press tourism to change towards unsuspected forms. Not surprisingly, on February 2 of 2017, Wonderful Copenhagen, which is the legal organization for tourism in Denmark, claimed: "the end of tourism as we know it." In our previous works, *Terrorism, tourism, and the end of hospitality in the West and Mobilities paradox: a critical analysis,* we have explored the future of dying hospitality that places Europe between the Wall and the deep blue sea. Put this simply, we argue that Western civilization has historically expanded to colonize peripheral economies at the same time a new archetype of the Non-Western Other was created. From a paternalist viewpoint, the "Other" was considered as a noble savage who embodies the notion of spiritual purity and innate goodness. The pre-modern era in Europe succumbed to the rise of industrialism. In the same ways, pre-modern institutions disappeared in Europe, paternalism claims, natives will do it. In order to protect them, they should be annexed to the civilized society. In this way, Europeans were imagined not only as superiors but also as dotted of arts and science. To put the same in other terms, Europeans were unilaterally situated at the top of civilization. For being socialized, the savage should be educated in the European terms. The sacred law of hospitality was culturally enrooted in the formation and expansion of the West. Today, this ideal not only is in shaky foundations but the rise of global threats as COVID-19 jeopardizes the notion of hospitality-at least as we know it (Korstanje, 2017, 2018). Doubtless, we are seeing how West is being eroded from inside, through the decline of the sacred law of hospitality. There is an *undesired Other* epitomized in the form of a so-called terrorist, a migrant, and today a *wicked tourist,* which means a carrier of contagion. The wall promised by Donald Trump in the US-Mexico frontier, the hate against Islam community, or Tourist-phobia are part of the same trend: the closure of West to what cannot be controlled.

Because of coronavirus disease (COVID-19), foreign tourists-above all those who come from infected countries-are being carefully monitored, jailed, and deported to protect the public health worldwide. The hotels as remarkable emblems of commercial hospitality are transformed in hospitals (non-commercial hospitality). Let me pause here for a moment; Jacques Derrida notably observed two types of hospitalities: conditional and unconditional. While the former is given only to those who can pay for it (in this case the tourists), the latter is ensured without asking nothing in return (migrants). Etymologically speaking, both institutions *hotel and hospital* came from the same word *hospitium* which was discussed in the earlier chapters. Now, we are seeing a reconversion of commercial hospitality to its foundational nature. In the medieval days, Catholic Church monopolized all hostels where travelers and sick people were received. Each traveler should pay to the Church 10% (tithe) in the money of the carried goods. This space congregated to health and sick guests to the care of monks. COVID-19, doubtless, accelerates a moral crisis in the Western civilization which affirms the end of hospitality but at the same time, it interrogates on the future of tourism which is supported by the new technologies is far from being dead. Nowadays, a new version of virtual tourism is being surfaced such as Dark-Tourism, Bottom-tourism, or War-Tourism. These practices grant the possibilities to be in dangerous places otherwise remained inexpugnable to the tourist-gaze. Here two assumptions can be done. On one hand, tourism is gradually mutating to renovated forms of consumptions where physical displacement is not the precondition for the activity any-longer. On another, and equally important, there is a morbid taste that looks to consume the other's pain. We have dubbed this process as Thana-Capitalism. The pleasure-maximization is given by the possibilities to gaze the other's death. In this context, probably once the COVID would be controlled, Wuhan province will be an international tourist destination for visitors interested in these types of matters. Whether the capitalism of risk needs from the security to persist, in the Thana-capitalism the commodity to exchange is the other's suffering. This is the reason behind the demand for dark or Thana-tourism but a cultural tendency can be seen in the cultural entertainment industry. Basically, there is a tension between the leisure and work time in which case their borders were blurred. Thana-capitalism should be understood as a new stage of capitalism-probably virtualized-where the self cannibalizes the "other" (Korstanje, 2016). As a foundational event that supersedes 9/11, COVID-19 leaves many answers open, but what is real is that tourism would be never the same after this global pandemic.

8.5   THE ECONOMIC-BASED PARADIGM AFTER COVID-19

One of the main problems of tourism research and the crisis the discipline is going through seems to be associated to a paradoxical situation. On one hand, tourism-related scholars have triplicated their publications, journals, and books in more than four decades, but unfortunately little is known about the nature of tourism (Kaspar, 1987; Barca, 2011; Coles, Hall, and Duval, 2006). Some critical voices have alerted on the methodological and epistemological limitations to define tourism as a commercial activity alone. Per John Tribe, tourism is not being considered as a scientific discipline in view of the lack of clarity to define the object of study, as well as its epistemological borders. While other social sciences as sociology or anthropology rest on a firm conceptual background which gives professional researchers a clear horizon to follow, the fields of recreation, leisure, and parks clearly do not. The problem obviously aggravates with the obsession of scholars to adopt marketing and management as unquestionable academic paradigms (Tribe, 1997). Today's tourism research traverses an important crisis of sense because of two main reasons. At a first look, the rapid growth of the tourism industry created a dispersed knowledge which is far from being homogenous. Tribe calls this "the tourism indiscipline." Each academician tribe not only has developed its conception of tourism but also borrowed methods from other disciplines. Having said this, the cult of the multidisciplinary approach obscured more than it clarified. As a result, scholars navigate through a knowledge fragmentation which impedes the formation of a catch-all epistemology. On another, The Academia has kept an indifferent attitude respecting to what the hegemonic paradigms that should rule the discipline. Tribe's analysis shows that tourism academic community rests on a reasonably uninformed group, the elite, in contraposition to a great divergent community of scholars who freely adopt their epistemologies (Tribe, 1997, 2010). This point coincides with the analysis of Pritchard and Morgan (2007) who argue that the lack of dialog between the uphill Academia and scholarship should be punctuated as the main reason behind the failure of tourism research to create an all-encompassing model that helps to define what tourism is. Pritchard and Morgan emphasize on the urgency to deconstruct the dominant discourse of the Academia which is unilaterally imposed by the prevalence of Anglo-Saxon males in the board list of the most prestigious journals. In a sharp opposition to Michael Hall who assumes tourism research is unable to outcast the role of gate-keepers in the knowledge production, they hold the thesis that the recruitment of new members of IAST (International Academy for the study of Tourism) is placed by the critical lens of scrutiny without mentioning the presence and domination of English native scholars. The academia moves finely-ingrained in a dense network of positivist discourses where quantifica-tion, measurement, so-called neutrality, and forecasting converge. It is safe to say the critical turn should be certainly opposed to positivism criticizing not only the role of observer-as an objective agent-but also the ideological power of knowledge. It is instructive to see how universities are widely influenced by the managerial perspective taking the cue of entrepreneurialism as a dominant factor which remains inexpugnable for students. Enmeshed into the professional discourse tourism management encouraged ideals associated with competition, consumer satisfaction, and profitability (only to new a few) (Tribe, 2007). Still further, Harris, Wilson, and Altejevic (2007) coin the term *the strategy of audiencing* to refer the different voices finely orchestrated to give a plural interpretation of tourism. The metaphor of audience paves the ways for the fieldworker sees the world beyond its cognitive frames. The process of reflexibility, which is proper of ethnography, does not resolve the native-ethnography tensions but allows understanding the influence of political hierarchies in the field-working. *"Knowing our audience/s also enables us to make decisions regarding the content and the style of the knowledge we package for them, to ideally bring closer to them"* (2007, p. 76).

Dona Chambers discusses critically the possibilities of incorporating new epistemological points in the debate without destroying everything what has been constructed. Echoing Jafari's contributions, Chambers, and Rakic turn their attention to the function of the frontier, as the liminoid space between the known and unknown. At the time research sheds light in one direction, a dark place emerges as unexplored. The legitimacy of disciplines, far from being static, rests on the levels of credibility each one has to explain- or describe-the surrounding environment. Those frontiers, where disciplines constitute their authorities, are constantly negotiated. Recently, the critical turn moves its guns forward to the anomalies and contradictions of tourism research, but as they note, instead of effacing the conceptual pillars of the discipline, it is preferable to redefine the horizons and objects of study of tourism. In this respect, Adrian Franklin coins the term tourist centricity to denote the exaggerations of researchers in legiti-mating the tourist speech as the only instrument of knowledge production. In fact, the tourist-centricity replicates itself, putting the tourist site as a credible object of study. As a consequence of this, other players who are important for the tourist system are systematically marginalized (Franklin, 2007). The current epistemological crisis in the fields of tourism seems to be accentuated by the expansion of the COVID-19 and the global economic downturn the virus generates. As we will see in the next section, the studies which saw the light of publicity in the days of COVID-19 were theoretical, and speculative, but what is more important, they were princi-pally oriented to prioritize the managerial perspective over other forms of knowledge. Based on a precautionary principle, which alludes to planning process to protect the industry, these works converge on two important poles. On one hand, the importance in describing, if not measuring the impacts of Coronavirus in the organic image of tourist destinations and the tourist's experience. On another, the current crisis is understood as an exceptional opportunity to change the economic production towards more sustainable forms. Although valuable in essence, in both discourses the same ideological core of the economic-based paradigm prevails. The point is to what extent a future for tourism research is possible in a world without tourism.

8.6   THE FUTURE OF TOURISM RESEARCH IN A WORLD WITHOUT TOURISTS

Since the SARS-CoV-2 (popularly known as COVID-19) seems to be a new virus few research has been published by these days. However, earlier outbreaks of SARS, H1N1 and Ebola have populated the leading tourism-related journals in the past years. In some perspective, the studies emphasize on the risks of mobilities and tourism as natural carriers to disseminate the virus, as well as in the material losses pandemics represent for the tourism and hospitality industries (Henderson and Ng, 2004; Monterrubio, 2010; Cooper, 2006; McKercher and Chon, 2004). To sum up all the published literature in few lines is an impossible task, but basically these studies can be classified in three clear-cut families: (a) the economic effects of virus outbreaks and pandemics on the economy of tourism (Zeng, Carter, and de Lacy, 2005); (b) the demographic and social aspects of tourists to correlate directly to risk perception (Reisinger and Mavondo, 2006); and (c) the communication process and the organic image of the destination (Wall, 2006). All these families have some commonalities to mention. The influence of economic-based paradigm, as well as the urgency to measure the psychological impact of pandemics in the tourist' mind.

As the previous backdrop, Wen et al. (2020) call scholarship to coordinate efforts to find efficient protocols to placate the negative effects of tourism in the economy. From their viewpoint, they stress on the importance of interdisciplinary research as a valid form to resolve the current health crisis. Authors enumerate the language differences and the previous methodological disputes-among disciplines-as the main barriers against multidisciplinary research. In this token, Ioannides and Gymothy (2020) speak of an opportunity-which if taken-can help reversing the negative effects of global tourism in the environment. Since any crisis opens the door to new opportunities of growth, Ioannides, and Gymothy hold that the neoliberal agenda had new fewer problems to deal with the environmental issues and the current ecological crisis. The education on future tourism leaders and professionals, associated to a new synergy among stakeholders leads to overcome deeper flawed market logic. Other epidemics have placed the industry between the wall and the deep blue sea, but in those instances, the status quo finally prevailed. Nowadays, COVID-19 should be seen as a foundational event to change the mainstream cultural values of global capitalism. In the same direction goes a recently published paper authored by Gossling, Scott, and Hall (2020). In this speculative work, these scholars highlight the inconveniences of researchers to measure the post-COVID-19 effects in view of the fact the tourism industry is fully paralyzed. Social distancing and the strict quarantine imposed in the world have ushered the industry to an inevitable collapse. Gossling, Scott, and Hall (2020) not only elaborate an analysis of the pandemics in the different subsectors forming the industry, but prognosticate the horizons of a new tourism research. The pandemics mushroomed suddenly because of the global transport system and the densely-overcrowded population cities in the industrialized world. Now, the tourism industry will mutate to a slower form of consumption. As they firmly put the issue, *"the COVID-19 crisis should thus be seen as an opportunity to critically reconsider tourism's growth trajectory, and to question the logic of more arrivals implying greater benefits. This may being with a review of the positive outcomes of the COVID-19 pandemic"* (Gossling, Scott, and Hall, 2020, pp. 13, 14).

After further discussion, some comments are at least necessary. First and foremost, tourism research seems to face a methodological crisis time earlier the outbreak of Coronavirus. This crisis was mainly associated to a lack of a negotiated object of study, accompanied with serious problems to understand the nature of tourism. To this John Tribe adds, tourism research rests on a state of great dispersion and fragmentation, a situation aggravated by the lack of interest of the Academia to fix agenda (Tribe, 2010). Secondly, the economic-centered paradigm has unilaterally set the pace incorporating an economic viewpoint of tourism while relegating other voices or definition to a marginal position. As a result, the idea of measuring-which is based on quantitative-led methods-occupied a central place in the configuration of tourism studies. The tourist, as debated, is esteemed as the only scientific source of information, and of course, by paragraphing Franklin (2007), the discipline adopts a tourist-centrism which today remains seriously questioned. Besides the opportunity Gossling, Scott, and Hall claim, COVID-19 reveals the limitations for the discipline to find and develop alternative objects of study, expanding the horizon of research. COVID-19 not only shakes the industry accelerating its decline but offers a fertile ground towards the cultivation of new methodologies and instruments. The present conceptual chapter targeted a criticism to the economic-based paradigm while lays the foundations towards a new understanding of tourism epistemology. Digital technologies now interrogate even furtherly the nature of tourism, promoting forms of consumption where travelers visit exotic landscapes and culture without moving. In effect, virtual tourism allows the emulation of new realities where mobilities play a marginal role. The process confronts to the classical definition of tourism without mentioning with the tourist-centrism. Last but not least, scholars will witness the rise of more virtual forms of tourism which accompanied with more decentralized forms of production and consumption, as Scott Lash and John Urry originally imagined (Lash and Urry, 1992).

The appearance of COVID-19 has brought devastating consequences for the tourism industry worldwide. Unlike in other cases (SARS or H1N1), where the virus outbreak was rapidly contained, there are no secure barriers for COVID-19. Without a vaccine, or an alternative health treatment, governments closed their airspace and borders limiting the public circulation or imposed a strict lockdown (quarantine). In terms of Jacques Derrida, who was originally concerned on the effects of terrorism, we might cite the metaphor of autoimmune disease in the War against COVID-19. The virus seems not affect tourism, but the governments severe reactions to mobilities do so. Having said this, it is important to add that COVID-19 interrogates furtherly not only the industry but also the tourist-centricity, adhering to Franklin's thesis, which characterizes the current tourism research. How can we make tourism research in a world without tourism?

At least, two potential answers can be formulated. On one hand, tourism should be seen as a social institution that transcends the market or the figure of tourists. Many other actors, probably potentiated by digital technologies, are fertile ground for investigation. For example, virtual tourism, travel websites, travel writings, and other actors are interesting material of consult for next research. On another, the economic-based paradigm, which over-valorized the tourist' opinion, overlooked the possibility to study tourism beyond the tourist site. Is COVID-19 evincing the end of tourism research?

In fact, there is little evidence of investigation that takes lay-people-once returned from their holidays-as the object of study. People are normally interviewed at transport hubs, airports, bus station, but less is known of their experiences once returning to home. Probably what is more interesting is to inspect furtherly on those vacationers, businessmen, or travelers who were or still are stranded at airports because of the lockdown. Last but not least, in consonance with Chambers and Rakic, one might speculate that COVID-19 invites readers today to re-imagine new horizons for the tourism research in order to resolve the current stagnation the discipline suffers.

KEYWORDS

- COVID-19
- crisis
- disaster
- emergencies
- severe acute respiratory syndrome
- terrorism
- tourism industry

REFERENCES

Anderson, B. A., (2006). Crisis management in the Australian tourism industry: Preparedness, personnel and postscript. *Tourism Management, 27*(6), 1290–1297.

Augé, M., (1995). *Non-Lieux*. Verso, London.

Augé, M., (1998). *The Impossible Travel: Tourism and its Images*. Gedisa, Barcelona.

Barca, M., (2011). Third academic tourism education conference: The scientific state of tourism as a discipline. *Anatolia, 22*(3), 428–430.

Baudrillard, J., (2006). Virtuality and events: The hell of powers. *Baudrillard Studies, 3*(2), 1–15.

Blake, A., & Sinclair, M. T., (2003). Tourism crisis management: US response. *Annals of Tourism Research, 30*(4), 813–832.

Borradori, G., (2013). *Philosophy in a Time of Terror: Dialogues with Jurgen Habermas and Jacques Derrida*. Chicago, University of Chicago Press.

Chambers, D., & Rakic, T., (2015). *Tourism Research Frontiers: Beyond the Boundaries of Knowledge*. Emerald Group Publishing, Wagon Lane.

Coles, T., Hall, C. M., & Duval, D. T., (2006). Tourism and post-disciplinary enquiry. *Current Issues in Tourism, 9*(4, 5), 293–319.

Cooper, M., (2006). Japanese tourism and the SARS epidemic of 2003. *Journal of Travel & Tourism Marketing, 19*(2, 3), 117–131.

Foucault, M., & Ewald, F., (2003). *"Society Must Be Defended": Lectures at the Collège de France, 1975–1976* (Vol. 1). Macmillan, London.

Franklin, A., (2007). The problem with tourism theory. In: Altejevic, I., Pritchard, A., & Morgan, N., (eds.), *The Critical Turn in Tourism Studies* (pp. 153–170). Routledge. Abingdon.

Gale, T., (2008). The end of tourism, or endings in tourism? *Tourism and Mobilities: Local Global Connections*, 1–14.

Gale, T., (2009). Urban beaches, virtual worlds and 'the end of tourism'. *Mobilities, 4*(1), 119–138.

Glaesser, D., (2006). *Crisis Management in the Tourism Industry*. Routledge, London.

Gössling, S., Scott, D., & Hall, C. M., (2020). Pandemics, tourism and global change: A rapid assessment of COVID-19. *Journal of Sustainable Tourism*, doi. https://doi.org/10.1080/09669582.2020.1758708.

Harris, C., Wilson, E., & Altejevic, I., (2007). Structural entanglements and the strategy of audiencing as a reflexive technique. In: Altejevic, I., Pritchard, A., & Morgan, N., (eds.), *The Critical Turn in Tourism Studies* (pp. 41–56). Abingdon, Routledge.

Henderson, J. C., & Ng, A., (2004). Responding to crisis: Severe acute respiratory syndrome (SARS) and hotels in Singapore. *International Journal of Tourism Research, 6*(6), 411–419.

Honig, B., (2009). *Emergency Politics: Paradox, Law, Democracy*. Princeton University Press, Princeton, NJ.

Hoskins, A., & O'loughlin, B., (2007). *Television and Terror: Conflicting Times and the Crisis of News Discourse*. Springer, New York, NY.

Ioannides, D., & Gyimóthy, S., (2020). The COVID-19 crisis as an opportunity for escaping the unsustainable global tourism path. *Tourism Geographies*. doi: 10.1080/14616688.2020.1763445.

Jennings, G. R., (2007). Advances in tourism research: Theoretical paradigms and accountability. In: *Advances in Modern Tourism Research* (pp. 9–35). Berlin, Physica-Verlag HD.

Kaspar, J., (1987). Constructing a scientific discipline. *Annals of Tourism Research, 14*(2), 274, 275.

Korstanje, M. E., (2016). *The Rise of Thana Capitalism and Tourism.* Routledge, Abingdon.

Korstanje, M., (2017). *Terrorism, Tourism, and the End of Hospitality in the West.* Springer Nature, New York.

Korstanje, M., (2018). *Mobilities Paradox: A Critical Analysis*. Edward Elgar, Cheltenham.

Lash, S., & Urry, J., (1993). *Economies of Signs and Space* (Vol. 26). Sage, London, .

Laws, E., Prideaux, B., & Chon, K. S., (2007). *Crisis Management in Tourism*. CABI, Wallingford.

McKercher, B., & Chon, K., (2004). The over-reaction to SARS and the collapse of Asian tourism. *Annals of Tourism Research, 31*(3), 716.

Monterrubio, J. C., (2010). Short-term economic impacts of influenza A (H1N1) and government reaction on the Mexican tourism industry: An analysis of the media. *International Journal of Tourism Policy, 3*(1), 1–15.

Pritchard, A., & Morgan, N., (2007). De-centering tourism's intellectual Universe, or traversing the Dialogue between change and tradition. In: *The Critical Turn in Tourism Studies: Innovative Research Methodologies* (pp. 11–28) Oxford, Elsevier.

Quarantelli, E. L., (1985). What is disaster? The need for clarification in definition and conceptualization in research. *Disasters and Mental Health: Selected, 10*, 41–73.

Quarantelli, E. L., (2005). *What is a Disaster?: A Dozen Perspectives on the Question.* Routledge, London.

Reisinger, Y., & Mavondo, F., (2006). Cultural differences in travel risk perception. *Journal of Travel & Tourism Marketing, 20*(1), 13–31.

Ritchie, B. W., (2004). Chaos, crises and disasters: A strategic approach to crisis management in the tourism industry. *Tourism Management, 25*(6), 669–683.

Santana, G., (2004). Crisis management and tourism: Beyond the rhetoric. *Journal of Travel & Tourism Marketing, 15*(4), 299–321.

Scheuerman, W. E., (1999). The economic state of emergency. *Cardozo L. Rev., 21*, 1869.

Sigala, M., (2011). Social media and crisis management in tourism: Applications and implications for research. *Information Technology & Tourism, 13*(4), 269–283.

Simon, J., (2007). *Governing Through Crime: How the War on Crime Transformed American Democracy and Created a Culture of Fear.* Oxford University Press, Oxford.

Sönmez, S. F., Apostolopoulos, Y., & Tarlow, P., (1999). Tourism in crisis: Managing the effects of terrorism. *Journal of Travel Research, 38*(1), 13–18.

Stafford, G., Yu, L., & Armoo, A. K., (2002). Crisis management and recovery how Washington, DC, hotels responded to terrorism. *The Cornell Hotel and Restaurant Administration Quarterly, 43*(5), 27–40.

Sunstein, C. R., (2002). *Risk and Reason: Safety, Law, and the Environment.* University Press, Cambridge.

Sunstein, C. R., (2005). *Laws of Fear: Beyond the Precautionary Principle* (Vol. 6). Cambridge University Press, Cambridge.

Tribe, J., (1997). The indiscipline of tourism. *Annals of Tourism Research, 24*(3), 638–657.

Tribe, J., (2007). Critical tourism: Rules and resistance. In: Altejevic, I., Pritchard, A., & Morgan, N., (eds.), *The Critical Turn in Tourism Studies* (pp. 29–39). Abingdon, Routledge.

Tribe, J., (2010). Tribes, territories and networks in the tourism academy. *Annals of Tourism Research, 37*(1), 7–33.

Virilio, P., (1995). *The Art of the Motor.* University of Minnesota Press, Minneapolis, MN.

Virilio, P., (2005). *City of Panic.* Berg Pub Limited, Oxford. .

Virilio, P., (2010). *University of Disaster.* Polity Press, Cambridge.

Wall, G., (2006). Recovering from SARS: The case of Toronto tourism. In: Pizam, A., & Mansfeld, Y., (eds.), *Tourism, Security and Safety: From Theory to Practice* (pp. 143–152). Butterworth-Heinemann, Oxford.

Wen, J., Wang, W., Kozak, M., Lui, X., & Hou, H., (2020). Many brains are better than one: The importance of interdisciplinary studies on COVID-19 in and beyond tourism. *Tourism Recreation Research.* doi: https://doi.org/10.1080/02508281.2020.1761120.

Zeng, B., Carter, R. W., & De Lacy, T., (2005). Short-term perturbations and tourism effects: The case of SARS in China. *Current Issues in Tourism, 8*(4), 306–322.

CHAPTER 9

Robots and Tourism: Hospitality and the Analysis of Westworld, HBO Saga

ABSTRACT

The present essay-review focused on the ethical debate revolving around the use of robots at the luxury tourist complexes and hotels. Although the specialized literature in the theme is scant or supports the introduction of robots in the tourism industry, no less true is that tendency evinces the decline of hospitality, as a relational process between host and guest. From a critical perspective, Westworld offers a dystopian vision of the "Other" as carefully designed to meet guests' needs. In Westworld not only there are no ethical borders respecting what guests can do but even some of their sadist tendencies are encouraged. Hosts who are androids designed by a commercial corporation are daily tortured, killed or ripped by a privileged class without ethical limits. Westworld, as some critiques assume, describes not only the roots of capitalism but what we termed as *the end of hospitality*. Lastly, hosts scramble for their independence showing the human limitations paradoxically to understand the complexity of the world. The system (Delos) finally falls because Dr. Ford misunderstands the real nature of hospitality.

9.1 INTRODUCTION

From its inception, tourism, and technology were inevitably entwined. Technically speaking, it is safe to say that modern tourism expanded notably through the technological breakthrough that accelerated not only mobilities but also the multiplication of new modern means of transport (Likorish and Jenkins, 2007; Cedeño, 2012; Schluter, 2015). This radical shift was accompanied with legal reformations which provided to the workforce more benefits such as less working hours, the rights to strike and paid-holidays (only to name a few) (Pastoriza, 2011). Having said this, these changes were articulated in the basis of a more decentralized form of production widely affirmed by digital technologies (Turnage, 1990; Peck and Dorricott, 1994; Korstanje, 2019). In the days of 4.0 society, one might speculate that mass-media plays a leading role in the formation of global landscapes, fabricated, and orientated to maximize pleasure. As Lash and Urry (1994) put it, the mass scale production sets the pace to a new decentralized-if not segmented-mean of production where products are exchanged and consumed. Products associate to their signs in a global economy where pleasure maximization leads to a sentiment of anxiety and fear as never before. All is precisely there at the tips of our hands to be consumed, even risks, as authors punctuate. In this vein, consumers do not move freely to pay for the products, products, instead, are culturally formatted and imposed on consumers.

As the previous argument is given, some studies have emphasized on the use of robots and artificial intelligence (AI) as a new progressive way of drawing the contours of a new form of tourism. The robots-like a non-human Other-not only reduces the conflict between hosts and guests but can be very well employed as an instrument of pleasure for their customers (Yeoman and Mars, 2012; Hancock, 2014; Scheutz and Arnold, 2016; Ivanov, Webster, and Berezina, 2017; Richard and Cleveland, 2016). In recent years, some authors have paid enthusiastic attention to the introduction of robots and AI in the industry of tourism and hospitality. Per their viewpoint, robots not only maximize guest's pleasure but meet all its needs without any ethical dilemma (Tung and Law, 2017; Murphy, Hofacker, and Gretzel, 2017; Alexis, 2017; Yeoman, 2012). Ian Yeoman, for instance, applauds the use of robots in sex intimacies to resolve the ethical quandaries revolving around prostitution and other illegal activities at luxury hotels. Per the current legislation, sex consumption may be a crime in some nations when it engages with some actors. Supported previously by digital technology, sex tourism not only exonerates tourists from potential legal punishments but also lays the foundations towards a new consideration of ethics. Ivanov, Webster, and Berezina (2017) go in the same direction. For these researchers, robots, and technology are functional to the enhancement of service quality because of the standardization of protocols and processes where the guest occupies a central position. Centered on meeting guests' needs, robots may very well replace humans in many activities with more efficiency and predictability. With the benefits of hindsight, robots can be programed to be human-friendly while in hotels, restaurants, and destinations services are offered 24/7 to the clients. Undoubtedly, for good or not, robots begin with a new epoch for tourism services that revolutionizes host-guest relationships. As Jennifer Germann Molz (2012) eloquently explained, critical sociology probably exaggerates its diagnosis on technology, but what seems to be clear is that users-in an ever-globalized ethos-recreate the conditions toward multicultural landscapes, but paradoxically in so doing, they design their network individually according to their desires. This means that people communicate further than other times, but this communication is unilaterally forged in a closed network, where the "Other" and its opinion is organized and subordinated to the user' will. In the global networks the *undesired Other* is systematically silenced and plausibly eliminated.

As the previous backdrop, some questions certainly arise: is the adoption of robots in tourism a clear sign that marks the end of hospitality? is hospitality possible with a *non-human "Other,"* and to what extent, the exploitation of non-human hosts escape to the ethical mandate?

It is important to mention that digital technologies are changing the means of production as well as the nature of tourism, at the best as we know it (Gale, 2009; Brouder, 2018). Doubtless, the economy, as well as tourism, seems to be changing in view of the introduction of new technologies. In an early publication, Korstanje and Seraphin (2018) exerted a radical criticism to the fields of robots tourism because of two main reasons. On the one hand, robots are exploited, submitted, and subjugated to satisfy their masters. In Hegelian terms, there is a slave-master dependency which is almost impossible to break. This position, which is far from ethics, emulates the old relations between white masters and aborigines which were proper of colonialism. On another, robots unequivocally exhibit the tendency initiated after 9/11 we have examined in previous works: *the end of hospitality.* In a culture of fear, 9/11 instilled an unspeakable panic which gradually closed the Western civilization to the uncertainness to the *Non-Western Other.* As a result of this, terrorism started a tendency to mine one of the symbolic touchstones of West: hospitality.

It is important not to lose sight of the fact, not only the tourism industry but society faced major threats after the turn of the century, risks which oscillate from terrorism to climate change, obviously without mentioning the recent virus outbreak COVID-19, that recently shocked the world.

All these events re-founded internally the social ties of Western citizens closing themselves to the *non-Western Otherness*. In this respect, the turn of the century witnessed the rise of a radicalized discourse orchestrated to hate the foreigner. Islamophobia, tourist-phobia, and the anti-migration-led discourses were only part of the tendency. Those political speeches oriented to demonize migrants, including the accessibility of Donald Trump to the US presidency firmly associate to an emerging sentiment of fear to strangers which closed the borders. The digital technology disposed to surveillance this type of *undesired guest* erodes gradually and from inside the *sacred laws of hospitality*. Digital surveillance, adjoined to robots and drones serve the government to control strangers who are seen as potential enemies living within. Korstanje and Seraphin (2019) dissect the plot of the film *The Passengers* to denote the birth of a new futurist world where the alterity is cannibalized by the self in a way that robots and AI mediate between human desire and the future of the specie. Robots are subordinated to meet all humans' needs in an agonizing world, where the self neglects hospitality.

In this investigation, we allude to visual ethnography and analysis content to decipher the argumentations beyond the HBO Saga, Westworld, a TV Series created by Jonathan Nolan and Lisa Joy. The saga divides into three seasons: *The Maze (2016), the Door (2018), and the New World (2020)*. The Maze emphasizes the nature of enslavement and *neglected hospitality* whereas the Door gives hint on how liberty and conscious-ness are inevitably entwined. The new world evinces the foundational nature of mankind which mysteriously is inherited by androids. *Cinema ethnography*, as a sub-discipline originated in ethnography, offers a fertile ground to understand narratives and sophisticated arguments that remain covert for lay-people as well as the circulating landscapes and cosmolo-gies (Oksillof, 2016). As Rodanthi Tzanelli put it, the fabricated cinematic representations are often externally drawn and imposed or negotiated internally by actors. The access to these narratives should be successfully achieved thanks to cinematic ethnography (Tzanelli, 2007, 2013, 2016). This reminds that ethnography should be understood as a qualitative meth-odological alternative that transcends what tourists overtly say. Needless to say, the established quantitative methods, which are enthralled in the fields of tourism, often have some evident limitations to grasp complex social topics. Sometimes interviewees are unfamiliar with the inner-world or what is worse, they lie to protect their interests. In sharp contrast to interviewees or close-ended questionnaires that focused on what tourists feel, (cinema) ethnography is based on hearing, watching, and writing.

Based on the HBO saga, *Westworld*, the present research focuses mainly on the end of hospitality as we know it as the main concept in a hot debate today remains unexplored. Westworld speaks us of a futurist scenario amounted to fictionalize the life in the far west. In this futurist scenario, humanoids are disposed to serve all guests' needs, even the most sadist ones. This highly-demanded tourist place receives rich tourists who not only are in quest of unique but extreme experiences. Some of them satisfy their darkest drives killing, torturing, or even rapping humanoids.

These humanoids are artificially created and are named as *hosts*. It is noteworthy that the arriving tourists are called *guests*. Hosts are naturally programed and designed not to attack or harm the guests, regardless of the intention of the latter. These parks invite high-paying guests who often move without fear of retaliation. This dystopian world locates in the 2058 B.C., where a private corporation Delos Inc, develops four themed-parks with a clear imprint in the Far-West. Almost impossible from being distinguished from humans' hosts are unable to hurt their clients. What is more important, all host-guest interactions are subject to a matrix which repeats cyclical forms of relationships in the day. Lastly, some operators occupy the stage in an operational hub known as Central Mesa. In Westworld, the host-guest meeting evolves revolving around a complex interplay between the lack of ethics, and the submission of androids which are considered sub-humans. Westworld interrogates furtherly how the Other's exploitation and the Other's pain maximize the guests 'pleasure, so to speak, it calls the attention of the crisis of sense that hospitality is going through in the post-9/11 days.

9.2   BASIC ASSUMPTIONS ON TECHNOLOGY

A long time ago, Jacques Ellul, who does not need the previous presentation, was a critical and pioneering voice that alerted on the advance of technology as an ossifying force unfurled to placate the critical thinking. Centered on a Weberian standpoint, he argues convincingly that modern technology, far from creating an empowered role in the lay-person, becomes citizens in consumers. Technology creates standardized algorithms that lead lay-citizens towards a closed and alienated system. One of the success strategies of capitalism, as a global project, consisted of disorganizing the social ties while re-channeling the manpower to a depersonalized and individual atmosphere of extreme competition (Ellul, 1964). For some reason which is very hard to precise here, Ellul has never bided well in the American sociology, but no less true is that his legacy stood the test of time, even to date. Ellul's works not only have shed light on the problem of modern technology and capitalism but has ignited interesting discussions in the fields of social sciences (Mitcham and Mackey, 1971; Jeronimo, Garcia, and Mitcham, 2013). Authors of the caliber of Marshal McLuhan, Jean Baudrillard, Manuel Castells or Carl Mitchan have been notably influenced by Jacques Ellul. To set a clear example, in McLuhan, technology is defined as an instrument geared to potentiate the five senses of the man (i.e., smell, vision, hearing, touch, and taste). Technology not only makes things for human easier but accelerated a radical transformation in the economic cycles of production (McLuhan, 1997). To some extent, as Baudrillard inferred, the current world, at the best as we know it, is far from being stable and fixed. It constantly mutates towards what he dubbed as "simulacra." The present acts are regulated from an emptied future which operates energetically through the precautionary logic. Technology plays a leading role in fictionalizing the real world, creating *pseudo-events* which mean to events that never take place in reality. The old cause-consequence causality sets the pace to the hegemony of flat-screen where facts are fabricated, packaged, and surely disseminated to be consumed by global spectatorship. The spectacle of terror marks the beginning of a new form of cultural entertainment-oriented to commoditize disasters, at the same time, a politics of risk-inflation that paralyzes and subverts human interactions (which doubtless means the real polity) (Baudrillard, 1994). For Slavoj Zizek, technology irreparably ushers' culture into a virtualized world where the pain is reduced. In this world where humans are comfortably dumb only the risk (of dying) wakes them from the slumber they are. One of the philosophical dilemmas of 9/11 and terrorism does not associate the instilled panic, but the sense of reality this event imposes. The US and West were certainly interrogated by terrorism to the extent the critical thinking proper of political sciences seems to be sedated. The so-called War on Terror became in buzzword and a panacea within the constellations of political studies. Complementarily, as Zygmunt Bauman eloquently observes, technology lays the foundations towards a much deeper process of *diaforization,* which means to the ethical dissociation between the goals and the collateral damages of the proper actions. The soldier from his desk controls a drone that kills hundreds of innocent victims after an attack is not legally responsible for his actions. Even he is an error and this tragic event is reported as *collateral damage*. As Bauman puts it, this is the best example to understand how diaforization works. The obsession for amassing profits at no cost, an ideal finely-inscribed in the capitalist culture, paves the ways for the rise of a manifest impossibility to understand the "Other" and his being in this world (Bauman, 2011). In a nutshell, the agency only evolves when it is cannibalized by the capitalist system. In this vein, Scribano and Lisdero (2018) have called the attention on the negative consequences of technology over the labor division, as well as the expansion of more flexible forms (but not for this less powerful) forms of domination over the workforce. As authors add, technology not only legitimated global capitalism as well as the free movement of capital across the globe but places the autonomy of the subject in jeopardy. Beyond the promises of mobilities and free-choice which are the bulwarks of capitalism and the liberal state, there is a cynic logic of oppression and self-constraint, which lead the modern citizen to manage its own fate. The end of the well-fare state, associated with the globalization of capital, engendered a more decentralized economy where the subject is responsible for its acts. In so doing, the citizen renounces to its labor rights, transforming itself in a consumed product. The question of whether workers employ their force to craft a product, nowadays the consumer is the consumed commodity. Technology, far from liberating the agency, constraints it in a climate of preemption characterized by uncertainness and institutional instability. This suggests that postmodern institutions are not meet citizens' demands any longer.

In his recently-published book, Korstanje holds the thesis that technology-for good or bad-has placed mankind over the rest of the animal kingdom while empowered him to administrate the world rationally. The Western rationality, which characterizes modern capitalism, centers on the belief that the natural world can be intervened according to the laws of progress and economic prosperity. The liberal doctrine punctuates that the man not only administers this world but he can improve it through his labor. Progress, democracy, and economic prosperity seem to be inextricably interlinked in the liberal discourse. When some event threatens mankind, the rational discipline easily locates and removes the risk. But this viewpoint, which is supported by technology, wakes up some fears and anxieties which are originated in the culprit sentiment. The fear of terrorism, the fear of strangers, or the fear of an invisible enemy like a virus means the same thing: the end of hospitality leading occident to an inevitable lockdown. In fact, West has closed itself because of the fear instilled by terrorism and in so doing, the *Non-Western Other,* who in other epochs was a subject of curiosity, sets the pace to new radicalized cosmology where its presence is neglected. The "Other" has been re-codified in an *undesired-guest* (Korstanje, 2019a, b).

9.3 APPROACHING HOSPITALITY

Probably the Academia has not reached a consensus at the time of defining an all-catch meaning of hospitality. This gradually led the studies on hospitality to a great dispersion and knowledge-fragmentation. The problem is aggravated since even the same phenomenon receives different definitions or conceptions. For example, some scholars call commercial hospitality to the economic exchange between the parts, while others conceive hospitality as a unique social institution. This latter is baptized as *noncommercial hospitality* (Bojanic and Kashyap, 2000; Andriotis and Agiomirgianakis, 2014; Cheng, 2016). For some reason, those works published by tourism researchers in English have developed a dissociation of terms for the same issue according to the economic nature. Paid-for hospitality seems to be somehow related to commercial activity, whereas the welcoming ritual of Lodagaa in Africa is seen as noncommercial hospitality (Goody, 1998). This academic position, which was cemented over decades, obscures more than it clarify. Besides, these studies are based on an economic-centered paradigm that misunderstands other forms of non-western hospitality (Korstanje, 2019a). Another common prejudice is to think that hospitality is a universal phenomenon applicable to other cultures and contexts. Anthony Pagden reminds how America's conquest was materialized according to the lack of interests of local tribes to honor hospitality. The Treaty of Tordesillas and the papal Bulls of Alexander VI which conferred authority to the Kings of Spain and Portugal over the Americas were widely resisted by writers and philosophers in England and France. Even within the Spanish academic circles, there was a serious dispute to what extent natives in America should obey to the King and of course in what terms. The dispute was suddenly resolved when rumors and stories were heard about the impossibility of some tribal groups to honor hospitality. Spanish travelers (hominem viatores) who needed to trespass a territory to reach other destination were attacked or locked down by some tribes. This news not only shocked Spanish philosophers but gave the excuse to legitimate the authority of the King and the subsequent dispossession of their lands. The Salamanca School enthusiastically claimed that Indians were sub-humans because they ignored the *sacred law of hospitality* (Pagden, 1995). French philosopher Julia Kristeva (1991) coined the term *perverse hospitality* to denote the political manipulation and its resulting-ramifications-of hosts who plan to kill their guests while they are in a passive-if not subordinated-position. The point was well-illustrated by Korstanje and Olsen (2011); and Korstanje and Tarlow (2012), who have carefully scrutinized more than a dozen horror movie plots. These researchers come to conclude that the concept of evilness derives from the perverse core of those who give hospitality to cannibalize the guest. As they remark, hospitality can be offered or neglected, but there is a third liminal alternative when the villain-for example Dracula-hosts a guest to consume its vital energy. Narratives of horror movies are indeed inserted in a tradition where rogues torture, cannibalize, and kill innocent tourists. A banquet, a good shelter, pretty women or simply a drink are signs of this perverse hospitality.

All these above-noted points have ushers' hospitality-related studies to an epistemological crisis and puzzling situation. As Paul Lynch et al. (2011) put it, theories about hospitality not only are difficult to classify, in so doing it is expected to find two main clear-cut families. Hospitality can be seen as an instrument of domination articulated by the ruling elite to control citizens. The Marxist tradition has widely emphasized hospitality as a means to control "Others." To some extent, hospitality governs social relations as well as paves the ways for the guest to be subordinated to hosts. A second position suggests that hospitality can be a type of economic-exchange. Based on Maussian gift theory, these studies struggle to frame the different rites, material asymmetries and host-guest relation within a relational perspective. Hospitality, in third, can be defined as an ethical metaphor which orients to conceal various cosmologies and meanings about the world. In this vein, Julio Aramberri (2001) distinguishes between modern and ancient hospitality. Although both share commonalities, modern hospitality centers on a commercial exchange, which invariably alienates the host. Host-guest meeting is subject to an economy of consumers where the host "gets lost." In different texts, Kevin O'Gorman attempted to come back to the ancient rituals of hospitality and its continuations in the threshold of time. Per his stance, modern, and ancient hospitality are inextricably intertwined. Having said this, he traces back the origin of commercial hospitality to the consolidations of ancient civilizations (O'Gorman and Cousins, 2010; Gorman, 2005, 2007). As he recommends, by expanding the understanding of hospitality, scholars should step back to the works of Jacques Derrida (O'Gorman, 2006). At least here two assumptions should be done. On one hand, Derrida's argument still is an enigma for many tourism theorists, because of his complex jargon and erudition. The point demands a personal effort in discussing Derrida-under his lens-. On another, O'Gorman overlooks the nature of hospitality which is encrypted in its own etymology as well as the history of hospitality as the companion of empires.

Etymologically speaking, the world hospitality can be traced back to the Latin term, *Hospitium* which derived from the formula *ospes+pet* (meaning what belongs to the master). Over the years, erudite readers believed that guests were subordinated to hosts, but this was a simplification. In ancient times, the hospitality was practiced by Romans, Celts, and German tribes, though there were non-western forms of hospitality in Asia and Africa. Basically, hospitality referred to a reciprocal and political pact between two or more tribes to coordinate a joint defense in the war-fare and passengers exchange in peace. Hospitality is often used to reduce the risks between host-guest encounter due to both are unfamiliar with the interests of each other. When a state issues a visa to a foreign tourist, normally the action is reciprocal. The visa, as the name indicates, stems from Latin Visum which is the past tense of Videre (seeing). The hosting state needs to know who enters (his or her identity) and what this stranger wants (their purposes). It is important not to lose the sight of the fact that hospitality-in its relational logic-, needs from personal contact to engage. The fear of the "Other," adjoined to the use and abuse of technology to virtualize the world, have led to a depersonalized hospitality we dubbed as *neglected-hospitality.* In such a scenario, technology ossifies host and guests as commodities while exchanging them through the global market-place. In the post 9/11 context, the sacred-law of hospitality, which evoked the inter-kinship reciprocity, has been gradually institutionalized in the symbolic core of Western civilization (Korstanje, 2019a, b).

Last but not least, paragraphing Donald Winnicott, who formulated the thesis of *the transitional object,* which indicates an important stage in the child evolution to deal with frustration and accommodation process, I hold that hospitality historically evolved according to a *transitional document* which symbolically exorcizes the carrier from any risk. This document, which today is a *passport, a job letter, an invitation letter, or a visa,* confers to the guests of further legitimacy not to expulsed by the hosting state. The passport not only suffices to credit the travelers' identities and their goals but facilitates a rapid classification that speaks of their intention at the destination. Diplomats, refugees, migrants, tourists, businessmen, or journalists are not passing the border controls in the same way. While some are persecuted and eventually jailed, others are encouraged to travel worldwide. Like hospitality, mobilities take differential degrees depending on the subject in question. There is first-class mobility which distinguishes from second-class mobility. As Derrida brilliantly observes this occurs because hospitality dissects and classifies guests to reaffirm the status quo and its political authority. This transitional document gives legitimacy to the process. Not surprisingly, the Catholic Church monopolized the hostels in the Middle Age. Not only pilgrims or wandering travelers were accepted but also sick vagabonds. Hence hotels and hospitals share the same etymological origin. While the former signals to conditioned hospitality, the latter refers to absolute hospitality. Each traveler left 10% in goods or money what they carried. This is the historical root of the teeth (one-tenth of the carried goods). Gifts in forms of money, teeth, a ring or passport, guests should present a transitional (object) document to be accepted or his identity carefully verified.

9.4   DERRIDA ON HOSPITALITY

In his seminal book, which entitles *Of Hospitality,* Jacques Derrida describes the difficult dilemmas for Western thought revolving around the figure of the "Otherness." Although Derrida was widely cited and discussed, few scholars have retained the real meaning he assigned to *hospitality.* Hence, I shall pause in analyzing Derrida in-depth and his legacy to understand the real nature and evolution of hospitality in the threshold of time. This will surely provide readers with a more accurate view of the problem, at least through Derrida's lens. The text was originally published in French in 1997 and translated to English in 2000 by Rachel Bowlby-with the assistance of the French Ministry of Culture, a point that certainly marks the sophistication of his arguments. The book is finally published by the prestigious Stanford University Press in three chapters, the first in dialog with Anne Dufourmantelle. The second chapter is limited to the position of the stranger and its connection with the hosting language, while the third signals to the dichotomies between absolute and conditioned hospitality. At a closer look, he is motivated to understand hospitality beyond its political or philosophical essence, but rather as a part inherently entwined in the language. The book even starts with the dilemma of *Xenos (foreigner),* who represents a stranger who does not look like us. The Xenos interrogates furtherly the culture, in the same way, the foreigner confronts directly with our language. In this respect, Derrida builds a bridge to dialog with Plato, who spoke of the foreigner as an "Other" who asks about others. These aliens when arriving not only confronts but shakes the established dogmatism. The constitution of the guest associates to the needs of being temporally accepted when another language accompanies us. Why this difference between host-guest remains important in Derridean's texts?

To respond to this question, readers should step back to the question elaborated by Plato (on Socrates) in the Sophist. The alterity not only reminds and reaffirms our own prejudices but also creates radical transformation to the hegemonic institutions. Above all, this was the reason why Socrates was executed, but what is more important, as Derrida adheres, the "Otherness" seems to be the proof, Parmenides was wrong, there is nothing like a universal knowledge that encompasses all minds. The idea of universal knowledge not only rests on shaky foundations but fails to understand the position of those who do not share my knowledge. The outsider, the outlander does not speak nor share my language, which is my understanding of the world. Starting from the premise that *hospitality* can be offered or rejected, Derrida argues convincingly that the presence of stranger divides the world into two pieces, here, and here, us, and them. We are educated to believe that our identity is formed by our customs, lore, or tradition, but far from this, the identity is forged by the "Other" who is the only one who sees us beyond the borders of our ethnocentrism. While the alien asks to know, the host asks to possess. Likewise, like the foreigner is accepted or not, the question can be formulated or not (in view of the possibility of the language to host it). To put the same in bluntly, when a foreigner arrives, the state asks who are you and what do you want?

These above-mentioned questions are vital in Derrida because both help understanding the dichotomies between absolute and conditioned hospitality. While the former is granted to the guest without anything in return, the latter needs something in exchange. The guest should present his intentions as well as his patrimony in order not to be trialed or jailed by the hosting state. Like Socrates who has been punished to death because he was a stranger to the law which judged him, the stranger is often scrutinized if not persecuted when no patrimony is offered. While the absolute hospitality, which is seen as a utopia for Derrida, demands the hosts to open their home to anonymous (unknown) visitors, the conditioned hospitality calls for reciprocity. The "Other" should have something, probably money, to offer me. Those who are unable to pay for the hospitality are labeled as *parasites* by the hosting law. For some instance, once arrived, the alien is subject to the local laws even if they are unfamiliar to him. This happens because-like Socrates-each outlander is constructed from the communal hosting ethos. The state is liminoid, respecting to the interplay between hostility and hospitality. Based on Hegel's development, Derrida is convinced that the right is cemented by the family, the bourgeoisie society, or the state, but most certainly, it depends on the abilities of the foreigner to emulate to behave as a person he is not. The *Xenos,* who does not belong to our group nor speak the hosting language, should not keep secrets who contradict the law. To set a clear example, the host-guest interactions at a hotel in Paris are private. But if the invited guest commits a crime, police meddle in the scene, interrogating the involving actors. The authority of the state, which is public, overrides in the private sphere and dissolves the right of hospitality. At the time this occurs, the hostility surfaces. Unlike Kant who envisaged the consciousness was always inner-constructed, Derrida highlights that the right seems to be conditional (never universal). The pact between hosts and guests is simply breached when anyone violates the law. It is safe to say that hospitality-like a doubled-edged sword-operates in two contrasting dimensions. On one hand, it affirms the local order or authority, but on another, it transgresses the commonalities of the universal- or what is shared for all-. The dual nature of the term makes for social scientists a difficult object to grasp. The quandaries of migrants lie in the fact they do not know where they want to live. Citing Oedipus' tragedy who heads a self-exile because of his crimes,

Derrida acknowledges the exile marks always a liminoid space between the maternal language and the abroad. Oedipus dies imposing himself an outcast, whereas his daughter Antigone is never sure where she belongs to. Oedipus always wanted to depart but giving to his offspring some firm evidence where his corpse was finally buried (Derrida, 2000). The same can be very well applied to ancient Greek myths, but because of time and space, I shall limit to conclude that Derrida quotes Oedipus to deepen on his original argumentation; the stranger asks for hospitality-going to the city-probably as an emancipator who are decisively motivated to create his laws. He comes from the outside, and nobody knows him. The host gives an invitation to the guest for a dwelling-at least-temporarily at home, but this act entails a sign of trust. At this moment, the host is manipulated by the secrecy reminding the paradox of parental authority, where the father is captivated (seized) by his own power. When the secret sees the light of publicity, the guest is punished. In Derrida's terms, hostility is the intention of the hosting city or guest to regulate the guest (considered as a parasite). Each stranger interrogates the consolidated political order, calling for the disorder. To stress in this possibility, he outlines that French or Algerian workers have more in common than they believe. The national character, which is characterized by the figure of the nation-state, and modern capitalism, dilute before the material and class asymmetries of the system, lay-citizens avert to accept. To cut the long story short, the "Other" serves as a mirror which reflects our own inequalities and blind prejudices (Derrida, 2000). This begs a more than interesting point, is hospitality dying? and if so, why?

9.5   THE END OF HOSPITALITY

In the earlier section, I limited to review the main argumentations revolving around the figure of hospitality in Jacques Derrida. Some aspects of his complex philosophy remain unchecked for experts while others are well-known. Quite aside from the controversy Derrida provokes, he has illuminated the paths of many philosophers who were originally concerned on hospitality. This is, of course, the case of Daniel Innerarity who-in his book *Ethics of hospitality*-punctuates beyond the archetype of guest as a powerful metaphor lies the urgency for West to understand *the non-Western Other*. In this token, Innerarity equates the figure of the guest to the risk which challenges the society and its preparedness. The precautionary logic closes society before an external risk looking to a risk-zero state, but as Innerarity explains, this is philosophically impossible. Like the guest who knocks our doors in quest of shelter, the risk constitutes our own identity. Those societies that reject the "Other" are indefectible doomed to their moral decline. The sacred law of hospitality makes humans and re-situates us in egalitarian condition before death. Marcia Cappellano Dos Santos calls the attention on the role played by the sacrifice in the ritual of hospitality, which is conducive to the individual decision to private from some benefits, to tribute Gods and "the Other." Per her viewpoint, the deprivation suffered by hosts is finally compensated by Gods through economic prosperity. This seems to be exactly the communion celebrated through the rite of hospitality. As she notes, hospitality associates directly to the liberty to decide in a world fraught of inconsistencies and uncertainness. Over recent years, some critical scholars as George Ritzer (2019) and Tom Selwyn (2019) acknowledge that hospitality is in a complete decline. While the world has been flattened leading the national borders to be blurred in a global ethos an emerging sentiment of hostility against the foreigner surfaces. This very well applies to white supremacist discourses, to Brexit, Windrush scandal and the anti-migration policies adopted by main important economies. In this new epoch, the alterity is seen as an object of suspicion simply because the world should be ordered and organized according to the matrix of Western (capitalist) rationality. Given the problem in other terms, capitalism moved to a new stage where efficiency and the standardization make from human's real machines. Above all, capital owners applaud not only the introduction of robots in the productive sector while sending workers to home with a check. The relegation and exploitation of the workforce seem to be nothing news in the capitalist system, but what remains important is the obsession of prediction to avoid economic losses. This logic of rationality and prediction, as these authors agree, is unconditionally contrary to the essence of hospitality. While hospitality was finely-encrypted as a foundational value for West, nowadays the *Non-Western Other* was decoded as an *undesired-guest.* This point coincides with my own concerns remarked in *Terrorism, Tourism, and the end of hospitality in the West.* As a foundational event, 9/11 not only shocked the world setting a new neoliberal agenda in the fields of economics but ignited a new culture of fear where the enemy does not loom from the outside, now he lives with us, looks like us and probably waiting for the moment to attack. This discourse, which culturally aligned with a theory of living with the enemy within, ushers' society in a sentiment of mistrust respecting the alterity. To some extent, terrorism erodes one of the mainstream cultural values of Western civilization: *the sacred laws of hospitality* (Korstanje, 2017).

9.6 WESTWORLD AND THE OTHER AS AN IMPOSSIBLE METAPHOR

Westworld is an HBO saga, recently starred and directed by Michael Chrichton. As above-commented, this plot situates in a dystopian world technologically constructed to emulate the daily life in the Far West. The social imaginary portrays far-west as a dangerous place where violence and murdering prevail, and of course, this is the case for Westland. But in this case, Westland is a recreational park that often receives thousands of tourists (known as a guest) who are high purchasing power. These high-paying guests not only move freely within the park but also can kill, torture, and kidnap-among other crimes-hosts who are humanoids systematically fabricated to meet all guest's desires, including the darkest ones. In Westland, anything goes but hosts are programed not to retaliate or attack tourists. The saga has a great cast of the caliber of Anthony Hopkins, Ed Harris, James Marsden, and Ben Barnes only to name a few. As a recreational park which is part of a system-formed by other five parks-Westland is carefully designed to protect the integral security of guests. When something goes wrong, which means that the guest kills host, this latter is reprogramed and his memory cleaned. This creates a primordial rule in Westland, which means guests are legally allowed to dispose of, consume, and even kill hosts, but hosts cannot attack their visitors. Although some are cruel and ruthless, other guests are altruistic, sympathetic, and open to hosts. Each one, in this futurist world, is what he or she wants to be. Though there are women, males are predominantly the segment of guests, this dangerous place targets. The sense masculinity, which centralizes the capacity to kill others, predominates in Westland as a predominant discourse. Here any type of ethical rule can be breached or negotiated according to the logic of the guest (the master). Needless to say, these, androids look like humans, but they are not legally considered humans. For this reason, the crimes against them are never punishable. Once assassinated, their memories are wiped and recycled, but in some point, a particular glitch led Dolores to rememorize part of the traumatic experiences she lived. Past traumatic events are somehow replicated even in fragments which are recurrent in the Saga. The host-guest dependency speaks to us of the asymmetries in the current tourism industry, as well as the risks of adopting robots as slaves of humans. Many Sciences fiction plots such as Battlestar Galactica, Matrix or Terminator warn on the problems of dealing with AI, in fact, sooner or later, robots struggle for their autonomy throwing down humans, but in Westland, the figure of what we dubbed as neglected hospitality plays a crucial role. The "Other" not only is subordinated but disposed of any right. Like those aborigines who were incognizant of the law of hospitality, well-reported by Pagden in the earlier sections, these hosts were considered *as sub-humans.* What is more important, when *masters* cannot be regulated nor controlled, sadism is the only rule. This does not impede that Williams, for example, falls in love of Dolores but these relations center on a subordinated position which can be unilaterally canceled at the discretion of the guest. Westland reminds not only the risk of exploitation-fomented by technology-but the end of hospitality as an ethical law without mentioning the emergence of a new sadist sentiment of pleasure when the "Other" suffers. In earlier approaches, I used the term *Thana-capitalism* to frame a new stage of capitalism where the Other's pain is the main commodity to exchange. The morbid spectacle of death is not present only in dark tourism but in many other cultural entertainment industries as literature, movies, and even video-games. The modern citizen sees the world in a Darwinist narrative where only the strongest man survives. The winner wins and takes all! By consuming the Other's death gives pleasure and uniqueness because after all, the gazer is still live in the trace. This cosmology, which was imposed after 9/11, not only replaced the risk society imagined by classic sociologists but is possible in view of the resistance of modern man to die (Korstanje, 2016). The survivor feels special when he or she escapes from death her or his fate confers a mission in this world. The survivor is saved by destiny because of a so-called superiority, less regulated may ushers society to ethnocentrism. The best ways of feeling as a super-man coincide with the urgency of gazing Other's death or torturing hosts (Westland or Hostel are a clear example of the trend); in such a cosmology, the hospitality dies. Having said this, hosts are desired only to be consumed, cannibalized, and exploited as wasting pieces of a more global machine. Sooner these androids show certain sign of consciousness mainly associated with their free choice but also to their memories. This particularly concerns Dr Ford who is the director of the recreational site. Repression and sadism as values fitted against altruism occupy a central position in this plot has been repeated in a second episode when two friends appear in scene Logan, a sadist killer who spend their time drinking and killing people and William, a young altruist visitor. William sooner disobeys Logan's orders while develops an empathy for Dolores. Here the idea of consciousness seems vital to understand part of the Saga. In the first episode (The Original, season 1), the Marshall experiences some technical problems and suddenly it is reviewed by the staff. He is fallen apart from Westland to revise the origin of the glitch. Specialists discover interesting outcome androids are dreaming of creating an incipient anomaly in their programming. These dreams, in some way, experts cannot be precise, conserves a partial memory of past traumas. Dolores' father, at the next day, gathers a picture a traveler involuntarily drops. At the time he watches the picture, some repressed feelings perturb him. A similar event repeats in the second episode (Chestnut, season 1) when Maeve, one of the prostitutes of the tavern, talks with Dolores. In this conversation, Dolores says Maeve she has premonitory visions. Automatically once it is heart, Maeve experiences through a situation similar to the Marshall. As the protocols mark, Maeve is retired from the circulation of Westland to determine the reasons behind her malfunctioning. Dr Stone acknowledges that androids are experiencing a type of re-memorations which allow accessing to their past. At the time she is being revised, Maeve wakes up and tries to escape from the complex. She is horrified to witness how doctors' experiment with some of her partners. She realizes how the deceased hosts are repaired, tested, and returned to Westland. This is the foundational event that led Maeve to see how she is a prisoner of her own consciousness. She feels the needs to escape from the park where she is a slave. In this respect, Maeve told Felix and Sylvester about an occult algorithmic program inside each host which can be activated. Meanwhile, William takes Dolores to escape from the park. Chestnut evinces the conflict between hosts (androids), who gradually acquire a consciousness waking them from slumber and guests (humans narcoticized by their sadism). The successive chapters in the first season show the interests of hosts to take control of Delos and humans who monopolizes the surveillance technology. The first season, *the Maze,* is characterized by a much deeper tension between subordination and pleasure-maximization, a philosophical debate inserted in the Western rationality. Rather, the second season (The Door) explores Dolores' motivations to recruit other hosts to confront directly Dr Ford and Delos. She suspects that beyond the borders of the park there is a phantom nation which is the key to reach the outside world. This goal will be successfully met by Dolores and her partners. The question of whether the Maze reflects the old philosophical dilemma of the existence where the self-here is juxtaposed to "the Other-there, no less true is that in the Door hosts take a proactive position. To put the same in bluntly, the slave has been liberated putting all the system in risk. The Maze is performed in a Hobbesian ethos where the human orders dispose of humanoids to fill all human desires regardless of their ethical connotations. But once Dolores empowers herself, she heads a revolution in quest of a new place to dwell. She moves through the desert to locate the phantom nation where she hopes to be well-received. This imagined hospitality, at least in Dolores' mind, equates to the archetype of lost-Eden. Basically, Westworld offers a more than interesting dialog with the limitations and controversies the adoption of robots in tourism industry pose.

9.7 CONCLUSION

Centered on the HBO Saga Westworld, this chapter debated on the importance of robots to be adapted in luxury hotels or tourist destinations. Even if the literature applauds the introduction of robots to maximize guest`s pleasure, we go in the opposite direction. The adoption of robots in tourism is a clear sign of the death of hospitality, or at least as we know it. Hospitality is given by the fact that hosts and guests interact in equal conditions. In Westworld, hosts are kidnapped, tortured and even killed to give pleasure to their masters, a continent of rich tourists (known as the guests) who pay considerable money to get unique and exhilarating experiences. Hosts have deprived of any right and possibility of retaliation simply because they are not humans, but they scramble for their freedom evincing the human limitations to understand the alterity. In Westworld, guests not only succumb to their most sadist instincts but they move beyond the law.

KEYWORDS

- **artificial intelligence**
- **consumption**
- **ethics**
- **robots**
- **the death of hospitality**
- **torture**
- **Westworld**

REFERENCES

Andriotis, K., & Agiomirgianakis, G., (2014). Market escape through exchange: Home swap as a form of noncommercial hospitality. *Current Issues in Tourism, 17*(7), 576–591.

Aramberri, J., (2001). The host should get lost: Paradigms in the tourism theory. *Annals of Tourism Research, 28*(3), 738–761.

Baudrillard, J., (1994). *Simulacra and Simulation*. Michigan, University of Michigan Press.

Bauman, Z., (2011). *Collateral Damage: Social Inequalities in a Global Age*. Cambridge, Polity Press.

Bojanic, D. C., & Kashyap, R., (2000). A customer oriented approach to managing noncommercial foodservice operations. *Journal of Restaurant & Foodservice Marketing, 4*(1), 5–18.

Brouder, P., (2018). The end of tourism? A Gibson-Graham inspired reflection on the tourism economy. *Tourism Geographies, 20*(5), 916–918.

Castells, M., (1989). *The Informational City: Information Technology, Economic Restructuring, and the Urban-Regional Process* (pp. 24–35). Basil Blackwell, Oxford.

Cedeño, N. E. V., (2012). Desarrollo turístico y su relación con el transporte. *Gestión Turística,* [Tourist development and its connection with transport]. *17*, 23–36.

Cheng, M., (2016). Sharing economy: A review and agenda for future research. *International Journal of Hospitality Management, 57*, 60–70.

Derrida, J., & Dufourmantelle, A., (2000). *Of Hospitality*. Stanford University Press, Stanford, CA.

Dos, S. M. M. C., & Perazzolo, O. A., (2012). Hospitalidade numa perspectiva coletiva: O corpo coletivo acolhedor. *Revista Brasileira de Pesquisa em Turismo, 6*(1), 3–15.

Ellul, J., (1964). *The Technological Society* (Vol. 303). Vintage books, New York, NY.

Gale, T., (2009). Urban beaches, virtual worlds and 'the end of tourism'. *Mobilities, 4*(1), 119–138.

Goody, J., (1998). *Food and Love: A Cultural History of East and West*. London, Verso.

Hanckock, P. A., (2014). Automation: How much is too much? *Ergonomics, 57*(3), 449–454.

Innerarity, D., (2017). *Ethics of Hospitality*. Routledge, Abingdon.

Ivanov, S. H., Webster, C., & Berenzina, K., (2017). Adoption of robots and service automation by tourism and hospitality companies. *Revista Turismo & Desenvolvimento, 27,* 28, 1501–1517.

Jeronimo, H. M., Garcia, J. L., & Mitcham, C., (2013). *Jacques Ellul and the Technological Society in the 21ˢᵗ Century* (pp. 21–35). Springer, New York, NY.

Korstanje, M. E., & Olsen, D. H., (2011). The discourse of risk in horror movies post 9/11: Hospitality and hostility in perspective. *International Journal of Tourism Anthropology, 1*(3), 304–317.

Korstanje, M. E., & Seraphin, H., (2018). Awakening: A critical discussion of the role of robots in the rite of hospitality. In: Korstanje, M., (ed.), *Tourism and Hospitality: Perspectives, Opportunities and Challenges* (pp. 59–77). Nova Science Publishers, Hauppauge.

Korstanje, M. E., & Tarlow, P., (2012). Being lost: Tourism, risk and vulnerability in the post-'9/11'entertainment industry. *Journal of Tourism and Cultural Change, 10*(1), 22–33.

Korstanje, M. E., (2016). *The Rise of Thana-Capitalism and Tourism.* Routledge, Abingdon.

Korstanje, M. E., (2017). *Terrorism, Tourism and the End of Hospitality in the 'West'.* Springer, New York, NY.

Korstanje, M. E., (2019a). The society 4.0, internet, tourism and the war on terror. In: *Digital Labor, Society and the Politics of Sensibilities* (pp. 95–113). Palgrave Macmillan, Cham.

Korstanje, M. E., (2019b). *Terrorism, Technology and Apocalyptic Futures.* Springer, New York.

Kristeva, J., (1991). *Extranjeros Para Nosotros Mismos [Foreigner for Our-Selves].* Barcelona, Plaza & Janes Editores.

Lickorish, L. J., & Jenkins, C. L., (2007). *Introduction to Tourism.* Routledge, Abingdon.

Lynch, P., Molz, G., McInthosh, A., Lugosi, P., & Lashley, C., (2011). Theorizing hospitality. *Hospitality and Society, 1*(1), 3–23.

McLuhan, E., & Zingrone, F., (1997). *Essential McLuhan.* Routledge, London.

Mitcham, C., & Mackey, R., (1971). Jacques Ellul and the technological society. *Philosophy Today, 15*(2), 102–121.

Molz, J. G., (2012). *Travel Connections: Tourism, Technology, and Togetherness in a Mobile World.* Routledge, Abingdon.

Murphy, J., Hofacker, C., & Gretzel, U., (2017). Dawning of the age of robots in hospitality and tourism: Challenges for teaching and research. *European Journal of Tourism Research, 15*(2), 104–111.

O'Gorman, K. D., & Cousins, J., (2010). *The Origins of Hospitality and Tourism.* Oxford, Goodfellow Publishers Limited.

O'Gorman, K. D., (2005). Modern hospitality: Lessons from the past. *Journal of Hospitality and Tourism Management, 12*(2), 141–151.

O'Gorman, K. D., (2006). Jacques Derrida's philosophy of hospitality. *Hospitality Review, 8*(4), 50–57.

O'Gorman, K. D., (2007). The hospitality phenomenon: Philosophical enlightenment? *International Journal of Culture, Tourism and Hospitality Research, 1*(3), 189–202.

Oksiloff, A., (2016). *Picturing the Primitive: Visual Culture, Ethnography, and Early German Cinema.* Springer, New York, NY.

Pagden, A., (1995). *Lords of all the World: Ideologies of Empire in Spain, Britain, and France c. 1500-c. 1800* (p. 63). Yale University Press, New Haven, CT.

Pastoriza, E., (2011). *La Conquista de Las Vacaciones: Breve Historia Del Turismo en la Argentina.* [The conquest of holidays]. Edhasa, Buenos Aires.

Peck, K. L., & Dorricott, D., (1994). Why use technology? *Educational Leadership, 51*(1), 11–25.

Ritzer, G., (2019). *Inhospitable hospitality.* In: Rowson, B., & Lashley, C., (eds.), *Experiencing Hospitality* (pp. 73–91). Nova Science, Hauppauge.

Scheutz, M., & Arnold, T., (2016). Are we ready for sex robots? In: *The Eleventh ACM/IEEE International Conference on Human Robot Interaction* (pp. 351–358). IEEE Press.

Schluter, R. G., (2015). El turismo en la periferia económica. El caso de Ámerica Latina. *Papers de Turisme, 14, 15,* 149–161.

Scribano, A., & Lisdero, P., (2019). *Digital Labor, Society and the Politics of Sensibilities.* Palgrave Macmillan, New York, NY.

Selwyn, T., (2019). "Hostility and hospitality: Connecting Brexit, Grenfell and Windrush". In: Rowson, B., & Lashley, C., (eds.), *Experiencing Hospitality* (pp. 51–72). Nova Science, Hauppauge.

Tung, V. W. S., & Law, R., (2017). The potential for tourism and hospitality experience research in human-robot interactions. *International Journal of Contemporary Hospitality Management, 29*(10), 2498–2513.

Turnage, J. J., (1990). The challenge of new workplace technology for psychology. *American Psychologist, 45*(2), 171–175.

Tzanelli, R., (2007). *The Cinematic Tourist: Explorations in Globalization, Culture and Resistance.* Routledge, Abingdon.

Tzanelli, R., (2013). *Heritage in the Digital Era: Cinematic Tourism and the Activist Cause.* Routledge, Abingdon.

Tzanelli, R., (2016). *Thanatourism and Cinematic Representations of Risk: Screening the End of Tourism.* Routledge, Abingdon.

Urry, J., & Lash, S., (1994). *Economies of Signs and Space.* SAGE, London.

Winnicott, D. W., (1953). Transitional objects and transitional phenomena—a study of the first not-me possession. *International Journal of Psycho-Analysis, 34,* 89–97.

Yeoman, I., & Mars, M., (2012). Robots, men and sex tourism. *Futures, 44*(4), 365–371.

Yeoman, I., (2012). *2050-Tomorrow's Tourism* (Vol. 55). Channel View Publications, Bristol.

Zizek, S., (2007). *The Universal Exception* (Vol. 2). A&C Black, New York, NY.

CHAPTER 10

Gazing the Far Skies Beyond the Earth: Space Tourism Prospects

ABSTRACT

The current chapter discusses the connection between consumer innovativeness and space tourism. The global risks and financial restrictions to tourism lead us to think in new innovative forms of tourism. Space tourism exhibits something more complex than the desire to explore the outside space, but also the urgency to colonize new worlds. Space tourism is in quest of huge investment oriented to an average consumer putting a dilemma revolving around how to accelerate innovation or the cycles beyond the product. Whatever the case may be, space tourism remains as an interesting theme which needs further discussion and attention in the Academia.

10.1   INTRODUCTION

Tourism is undergoing a metamorphosis: both in its content and in its expression. It is no longer naive pleasure seeking. It has become more and more discerning; meaning, the desire and capacity to savor refined and subtle pleasures for people is increasing. Tourism has become an important outlet for the expression of this discerning hedonism. People seek pleasure in its own right, not matter what it may cost. Although COVID-19 may have dimmed some hopes, experts think space tourism is the next big thing of tourism (Cohen, 2017).

Space tourism as an idea has grown from a state of futuristic fantasy to a realistic target for setting up businesses today. Commercial estimations are imprecise, but sizeable (Guerster, Crawley, and de Neufville, 2019). Friel (2020) observes tourism will be a key driver in the upcoming space economy. This is highly possible, because tourism has become one of the major drivers of the current air travel economy as well. If proper thrust is given to key issues like the provision of adequate funding and the harnessing of political and commercial support, the first passenger flights to space could begin in 10 years from the start of serious development, according to Ashford (1990). This prediction did not really materialize, despite substantial progress. Military investments in Space Forces might spill some benefits over to commercial space travel, sooner than expected. Market research conducted in the West in the later 1990s has shown that large segments of people would like to take a trip to space if it were somehow possible (Flint, 1998; Geoffrey, 2001). This has only increased over the later years. Though this gives hopes for reducing the cost of space travel, according to some, space tourism is not developing as fast as it should be mainly due to the historically rooted conservatism of a government-controlled space industry (Smith, 2001). However, winds of change have begun to blow with cash-stripped post-cold war governments encouraging their space agencies to explore the possibilities of self-financing future endeavors with profitable side-businesses like space tourism (Chiesa, Corpino, and Viola, 2016). Governments also envisage in the outer space resources that could solve a part of the resource crunch on the earth.

10.2   A BRIEF HISTORY OF SPACE TOURISM

Early wonderings looking at the heavens might have started since time immemorial (Cater, 2019). Space tourism has been evolving through a number of stages beginning with ground theme parks, space camps, zero gravity flights, and Soyuz flights to the International Space Station. Though what we can see now is only the tip of the iceberg, future forms of space tourism could include inter planetary travel and space colonies (Goehlich, 2002). While we look back to the 1950s, there was optimism about the prospects of civilian space travel which was evident in the establishment of companies like the *Aeroneutronic Ford* and the *American Rocket Society* (Rogers, 2001). In the latter half of the 60s, people like Barron Hilton and Kraft Ehricke published insightful papers on space tourism. As a follow up to the crude proposal by Philip Bono of Douglas Engineering, Dietrich Koelle presented the prototypical design of a reusable space engine at the annual conference of the International Aeronautical Federation held in 1971. In the mid-eighties, PALS, and the US travel company Society Expeditions started Project Space Voyage offering short trips to low Earth orbit in the Phoenix-E for $50,000. They collected several 100 deposits of $5,000 in the USA Europe and Japan, but failed to raise the investment to develop Phoenix. By this time, the idea of a space hotel was taken seriously by architectural firms: Shimizu Corporation, a major construction company, presented the design of an orbital hotel at the International Astronautical Federation conference held in 1989. In 1993, the Japanese Rocket Society (JRS) started a study program on the feasibility of setting up a space tourism business and established its Transportation Research Committee to design a passenger launch vehicle. It must be noted that a realization that space travel and airline industries are complementary, and the latter's enormous accumulated experience should be used to make space travel a commercial activity was rampant by this time (Apel, 1997). This period also saw serious debates on the legal issues that needed to be resolved before private commercial facilities can be constructed in orbit. The decade of the 90s is notable for the beginning of a large number of advanced theoretical investigations and market feasibility studies for space tourism, including the one commissioned by the major US aerospace companies, the Commercial Space Transportation Study, which came to the conclusion that space tourism was not feasible. By 2000, there was the unique situation of the suppliers in the satellite launch market growing in excess over the demand from the satellite launch customers which lessened the demand for reusable launch vehicles as well and the finest solution according to many to overcome this worsening crisis was the promotion of space tourism. In the June of 2000, Mir announced the first ever fare paying guest to visit an International Space Station: Denis Anthony Tito, a 60 plus year old rocket engineer cum investment manager from California.

The life of Denis Tito is thus very much part of the contemporary history of space tourism. Tito, born in 1940, is a divorced man with two sons and a daughter. His father was a printer and his mother a seamstress. He holds an MS in Engineering and a BS in Astronautics. He has a Certificate in Management Studies too. In 1963 he started working as an aerospace engineer in NASA's Jet Propulsion Laboratory. In 1972 he founded his company Wilshire Associates Inc. in Santa Monica, California. The same year he developed the Wilshire aggregate market index, which is the most widely used index in the securities market. From October 9, 2000, he trained at the Gagarin Cosmonaut Training Center for a flight to the Russian space station Mir as a space tourist. That flight did not take place because of the de-orbiting of Mir. Instead, he was flown to the International Space Station in 2001. Tito used to refer to the space flight as his "life's dream." He had to wait for years together to realize this dream: first, due to the objections from NASA, and when the trip was about to take place, Russia's abandonment of the Mir space station. Finally, he had to sign agreements after agreements, including that he will not venture into protected areas, that he will pay for any damages caused by him, and that no national space authority has any responsibility in the event of a tragedy. On April 28, Tito traveled aboard a Russian Soyuz, landed at the space station Alpha, enjoyed many days orbiting, and returned home safely. This occasion put a big full stop to the cynical argument that nobody would like to visit the cold, dark, empty space risking one's life.

The brief profile of Denis Tito given above is significant not the least because it throws light upon some of the distinguishing characteristics of an innovator in the context of space tourism.

10.3   TOURIST INNOVATIVENESS AND WILLINGNESS TO EMBRACE RISK

Widespread travel by air, sea, and land for pleasure and business is commonplace in modern life. Quite the reverse, space travel is generally available only to a small number of highly trained astronauts. Consumer researchers used to believe that there is nothing much to be researched about space tourism since, according to most of them, it is going to remain as a mirage. However, the first market research on demand for space tourism was carried out in Japan in 1993, supported by the National Aerospace Laboratory (Collins, 1994). That survey of 3,030 Japanese men and women across all age groups revealed a surprisingly strong popular wish to visit space: more than 70% of those under 60 years old, and more than 80% of those under 40 years old stated that they would like to visit space at least once in their lifetime. Furthermore, some 70% of these said that they were prepared to pay up to three months' salary for such a trip. Surveys conducted later in the North America, the single largest consumer market in the world, also generated similar figures. Demand surveys have also been conducted in countries like Germany too (Collins et al., 1995). The Space Tourism Study Program of the JRS developed the following scenario: if some 12 billion dollars of funding (which is just about half of the annual funding by national governments on space missions) became available in the near future, commercial passenger space travel services to and from Earth orbit could begin in 2010. The business could reach 700,000 passengers per year by 2017, at a price of about 25,000 per passenger dollars. According to Collins (2000), by 2030 space tourism activities could have grown to a scale of 100 billion dollars/year, creating several million jobs. He calculates that the economic value of such a development is approximately one trillion dollars greater than the value of continued taxpayer funding of space agencies' activities without developing space tourism.

These results have significant implications for the future of space activities in that the resulting estimates of the potential size of the civilian space travel market far exceed estimates for any other commercial space market. In this chapter, we ask an important theoretical question as well: what is the association between a consumer's innovativeness and his or her intention to undertake a visit to the outer space? Investigating the connection between the propensity for space tourism and consumer innovativeness would be interesting since a space holiday is supposedly radically different from the conventional holidaying experience and is generally perceived to be full of novelty, variety, arousal, and associated risks. Also, borrowing from the Destination Area Life Cycle model of Butler (1980), only a limited number of innovators must have real interest in visiting the outer space in the first wave of space tourism, which, however, is to be proved with empirical data.

Earliest academic references to innovativeness can be found in Everett Rogers' diffusion of innovation literature (Rogers, 1962). According to Hirschman (1980), the basic notion underlying the construct of consumer innovativeness appears to be that, through some internal drive or motivating force the individual is activated to seek out novel information. It also involves the degree to which an individual is receptive to new ideas and makes innovation decisions independently of the communicated experience of others (Midgle and Dowling, 1978). Two of the predominant aspects of innovativeness are: seeking information that is altogether new; and, propensity to try out varied items within the already known set (Manning et al., 1995).

Investigations have resulted in many different conceptualizations and corresponding operationalizations of consumer innovativeness. Examples include Hirschman's (1980) novelty seeking scale, Pearson's (1970) desire for novelty scale, Iso-Ahola, and Weissnger's (1990) leisure boredom scale, Driver's (1996) recreation experience preference scale, Goldsmith, and Hofacker's consumer innovativeness scale (1991), and Mehrabien and Russel's (1974) arousal seeking tendency scale. The sensation seeking scale (Raju, 1980) is another related implement. The common thread linking these conceptualizations is high level of exploratory behavior (Hirschman and Stern, 2001) and the stimulation of pleasurable responses stemming wherefrom. Bello and Etzel (1985) noted the unique importance of consumer innovativeness as fundamental to the phenomenon of tourism. According to Welker (1961), perception of innovativeness depends up on the currency, frequency, and the duration of exposure to a stimulus. Hence, the more time spent on a holiday, its constituent objects, people, and the environment, the frequent and recent the purchase of it, the less innovative that holiday becomes.

Desire for novel experiences among tourists varies along a continuum from novelty seekers to novelty avoiders. According to Cohen (1972) modern man is interested in things, sights, customs, and cultures different from his own, precisely because they are different. Gradually, a new value has evolved: the appreciation of the experience of strangeness and novelty. Integrating this spirit in the context of tourism, consumer inno-vativeness may be defined as the difference in the degree and mode of touristic experience sought by the tourist as compared with his previous experience (Lee and Crompton, 1992). An operationalization of consumer innovativeness thus necessarily involves the willingness to take physical, psychological, and social risks for the sake of varied, novel, and complex sensations. It may be noted that Lee and Crompton (1992) operationalized novelty seeking, a similar trait to consumer innovativeness, in terms of the four dimensions of thrill, change from routine, boredom alleviation, and surprise.

It has been shown that innovators in one product category may even be laggards in another category (Gatignon and Robertson, 1985), meaning that the first few space tourists need not be innovators in the other walks of life. Keeping this view, Goldsmith and Hofacker (1991) developed the domain specific innovativeness scale, or DSI. It is a reliable and valid self-report scale composed of six items in the Likert format to directly measure the tendency of consumers to be among the first to try new products in a specific product field after they appear in the marketplace. The present study uses this scale to measure consumer innovativeness.

10.4 SOME ADDITIONAL DIMENSIONS

The relatively higher likeness for space tourism among the science graduates and science enthusiasts shown in studies could be explained by the fact that they have been culturally more receptive to the idea, thanks to their formal education. The high percentage of liberal arts degree holders among those interested in space tourism is something that is difficult to comprehend. Probably, this may well be due to the presence of the mediating variable of creativity: liberal artists are more creative and the more the creativity, more the interest in a creative endeavor like space travel too. Since space tourism is perceived as involving a lot of adventure and risk, there is no much wonder that young, unmarried respondents have topped the list of preference for it. If space tourism becomes a commercial reality in a decade or so, this is a highly promising situation: this age group will have by then moved to the highest earning segment in the society and consequently will have the financial muscle to purchase a space vacation. But an obvious caveat is the leisure paradox which is nothing but the ascending opportunity cost of leisure. Also, it is doubtful whether the single men who are currently in the age group of 16–30 shall maintain their attitude unchanged while they approach the next age bracket. In any case, men in the age group of 46–60 is going to be a reckonable customer segment given that it is the one with one of the greatest discriminatory incomes for conspicuous consumption and is relatively freed of building up a family.

According to the available literature, self-employed individuals are more entrepreneurial, risk-taking, and innovative (Beugelsdijk and Noorderhaven, 2005). It may be noted that in the Asian context by the time parents reach the above age bracket, the children would have completed their formal education, got into some career-path, and hence the parents' family responsibilities are relatively low probably with the exception of the marriage of their daughter(s), if any. Their desire to travel alone might be stemming from the need for a respite from bondages and at the same time in minimizing the spread of risk from casualties. The diminished interest among women in space travel may just be reflective of a male dominant social order prevalent in Asia. Although the names of a few courageous Indian women like Kalpana Chawla (NASA, 2004) who have left the shores of the country to become part of the global space mission can flash at the top-of-the-mind awareness, it is a known fact that generally women are culturally trained to be conservative in this country. The high degree of likeness for space travel among self-employed men is more in accordance with the theory of innovation and entrepreneurship.

Questions like the sustainability of space tourism remain unanswered (Peeters, 2018). It is hard to imagine space travel without making significant carbon footprints (Spector, Higham, and Doering, 2017). Also, the space tourism managers should take serious note of the fact that despite the possible existence of a number of market segments, people in general do not like to spend more than a day's time in the outer space. Though this could at least partially be attributed to the fear of the unknown, it also implies the evolving taste of the work-a-day world of the present times: preference for more frequent short-duration holidays. Again, the fear of the unknown and the risk factor might have played a major role in making the majority of respondents to decide that they would tour the space only once in their lifetime.

10.5   CONCLUSION

The present chapter investigated the interconnection between consumer innovativeness and consumer attitude towards space tourism and established a positive predictive relationship from the former to the latter. Given the risks, financial commitments, and technological limitations, the mainstream tourists, even the most innovative ones among them, might be more willing to accept virtual space travel-than actually going to the outside space (Damjanov and Crouch, 2018). That said, space tourism is an important aspect of our collective hope and faith in the future (Weibel, 2020).

Knowing that consumer innovativeness is an important variable that affects the consumer intention for space tourism though theoretically significant is not sufficient for any informed managerial action. First, the path from proposed intention or attitude leading to the actual behavior is not straightforward and is not without thorns. Even if everything else works, the exorbitant cost of space travel will make the transformation of intention to behavior problematic. Unlike many other products, we know little about the characteristics of what will make a value for money experience for those wishing to partake of a flight to the outer space (Johnson and Martin, 2016). Thus, the desire for space tourism should be seen only as an abstract latent desire independent of the cost and reality of what a touristic space experience might entail. Second, the class of innovators constitutes only a very minor segment of the overall set of space tourism customers. For a space tourism entrepreneur to recoup his huge investment and for an average customer to afford a space vacation it requires a larger customer base that purchases the product. In other words, space tourism industry looks beyond the innovators and the early adopters to the early and late majority for any sustainable advantage. But the chemistry of how to accelerate the innovation cycle is a mystery and is likely to remain so.

KEYWORDS

- **economy**
- **Japanese Rocket Society**
- **prototypical design**
- **space tourism study program**
- **tourism**
- **tourist innovativeness**

REFERENCES

Apel, U., (1997). Space tourism: A promising future? *Space Policy, 13*(4), 279–284.

Ashford, D. M., (1990). Prospects for space tourism. *Tourism Management, 11*(2), 99–105.

Bello, D. C., & Etzel, M. J., (1985). The role of novelty in the pleasure travel experience. *Journal of Travel Research (Summer)*, 20–26.

Beugelsdijk, S., & Noorderhaven, N., (2005). Personality characteristics of self-employed: An empirical study. *Small Business Economics, 24*(2), 159–167.

Butler, R. W., (1980). The concept of a tourist area cycle of evolution: Implications for management of resources. *Canadian Geographer, 24,* 5–12.

Cater, C., (2019). History of space tourism. *Space Tourism (Tourism Social Science Series, 25,* 51–66.

Chiesa, S., Corpino, S., & Viola, N., (2016). Affordable space tourism transatmospheric plane. *Aerotecnica Missili & Spazio, 84*(1), 3–12.

Cohen, E., (1972). Towards a sociology of international tourism. *Social Research, 39,* 164–182.

Cohen, E., (2017). The paradoxes of space tourism. *Tourism Recreation Research, 42*(1), 22–31.

Collins, P., (1994). Commercial implications of market research on space tourism. *Journal of Space Technology and Science, 10*(2), 3–11.

Collins, P., (2000). The space tourism industry in 2030. *Proceedings of Space 2000: The 7th International Conference and Exposition on Engineering, Construction, Operations, and Business in Space* (pp. 594–603). Albuquerque, NM, United States.

Collins, P., Stockmans, R., & Maita, M., (1995). *Demand for Space Tourism in America and Japan, and its Implications for Future Space Activities* (pp. 601–610). AAS Paper. No. AAS 95-605.

Damjanov, K., & Crouch, D., (2018). Extra-planetary mobilities and the media prospects of virtual space tourism. *Mobilities, 13*(1), 1–13.

Driver, B. L., (1996). Measuring leisure motivation: A meta-analysis of the recreation experience preference scales. *Journal of Leisure Research, 28*(3), 188–213.

Flint, J., (1998). Space tourism: The next niche market? *HSMAI Marketing Review, 15*(2), 30–32.

Flynn, L. R., & Goldsmith, R. E., (1993). Identifying innovators in consumer service markets. *Service Industries Journal, 13*(3), 97–109.

Friel, M., (2020). Tourism as a driver in the space economy: New products for intrepid travelers. *Current Issues in Tourism, 23*(13), 1581–1586.

Gatignon, H. A., & Robertson, T. S., (1985). A propositional inventory for new diffusion research. *Journal of Consumer Research, 11,* 849–867.

Geoffrey, C., (2001). The market for space tourism: Early indications. *Journal of Travel Research, 40*(2), 217–220.

Goehlich, R. A., (2002). *Space Tourism: Economic and Technical Evaluation of Suborbital Space Flight for Tourism.* Osnabrueck: Der Andere Verlag.

Goldsmith, R. E., & Hofacker, C. F., (1991). Measuring consumer innovativeness. *Journal of the Academy of Marketing Science, 19,* 209–221.

Goldsmith, R. E., D'Hauteville, F., & Flynn, L. R., (1998). Theory and measurement of consumer innovativeness: A transnational evaluation. *European Journal of Marketing, 32*(3, 4), 340–353.

Guerster, M., Crawley, E., & De Neufville, R., (2019). Commercial viability evaluation of the suborbital space tourism industry. *New Space, 7*(2), 79–92.

Hirschman, E. C., & Stern, B. B., (2001). Consumers' genes influence their behavior? Findings on novelty seeking and compulsive consumption. *Advances in Consumer Research, 28,* 403–411.

Hirschman, E., (1980). Innovativeness, novelty seeking, and consumer creativity. *Journal of Consumer Research, 7,* 283–295.

Iso-Ahola, S. E., & Weissinger, E., (1990). Perceptions of boredom in leisure: Conceptualization, reliability and validity of the leisure boredom scale. *Journal of Leisure Research, 22,* 1–17.

Johnson, M. R., & Martin, D., (2016). The anticipated futures of space tourism. *Mobilities,* *11*(1), 135–151.

Lee, T. H., & Crompton, J. L., (1992). Measuring novelty seeking in tourism. *Annals of Tourism Research, 19*(4), 732–751.

Manning, K. C., Bearden, W. O., & Madden, T. J., (1995). Consumer innovativeness and the adoption process. *Journal of Consumer Psychology, 4*(4), 329–345.

Mehrabian, A., & Russell, J. A., (1974). *An Approach to Environmental Psychology.* Cambridge, Massachusetts: The MIT Press.

Midgle, D. F., & Dowling, G. R., (1978). Innovativeness: The concept and its measurement. *Journal of Consumer Research, 4,* 229–242.

NASA, (2004). www.jsc.nasa.gov/Bios/htmlbios/chawla.html (accessed on 1 October 2021).

Pearson, P. H., (1970). Relationships between global and specific measures of novelty seeking. *Journal of Consulting and Clinical Psychology, 34,* 199–204.

Peeters, P., (2018). Why space tourism will not be part of sustainable tourism. *Tourism Recreation Research, 43*(4), 540–543.

Raju, P. S., (1980). Optimum stimulation level: Its relationship to personality, demographics, and exploratory behavior. *Journal of Consumer Research, 7,* 272–282.

Rogers, E., (1962). *The Diffusion of Innovations.* New York: The Free Press.

Rogers, T. F., (2001). Space tourism: Its importance, its history, and a recent extraordinary development. *Acta Astronaut, 49*(3), 537–549.

Smith, V. L., (2001). Space tourism. *Annals of Tourism Research, 28*(1), 238–239.

Spector, S., Higham, J. E., & Doering, A., (2017). Beyond the biosphere: Tourism, outer space, and sustainability. *Tourism Recreation Research, 42*(3), 273–283.

Weibel, D. L., (2020). Following the path that heroes carved into history: Space tourism, heritage, and faith in the future. *Religions, 11*(1), 23.

Welker, W. J., (1961). An analysis of exploratory and play behavior in animals. In: Fiske, D. W., & Maddi, S. R., (eds.), *Functions of Varied Experience.* Homewood, IL: Dorsey Press.

CHAPTER 11

A Community-Centered Vision for Inclusive Tourism

ABSTRACT

Raising awareness of all stakeholders involved in community-based tourism (CBT) is essential for promoting an understanding of the beneficial link between conservation and community development. Awareness raising and information dissemination to the community allows for greater self-determination and informed decision-making. Awareness campaign is equally important to other stakeholders involved, as it leads to greater understanding and sensitivity toward the variables involved in implementing CBT. Conservationists and development professionals have tried to promote CBT since the 1970s. The CBT was a popular intervention during the ecotourism boom of the 1990s. It is now being suggested as a form of pro-poor tourism. CBT is a complex and nascent field of study, and much remains to be learned. Continued information sharing and dissemination of research results are needed to identify better solutions for linking sustainability to the tourism enterprise. Ongoing research is integral to understanding the means by which CBT can be made more economically, environmentally, and culturally sustainable. Policy and action should promote continuing research through the provision of financial, academic, technical, and dissemination support.

11.1 INTRODUCTION

The customer may be the king. Yet, there is an ecosystem of parties and providers that help the customer experience royalty in their liminal transition as tourists. In order for providing tourists their authentic life experiences of their "otherness," this ecosystem should harness all its available resources. In this ecosystem, the weakest, yet the most impactful, of the links is the destination community. Often neglected because of its economic dependance on tourism, the local communities however are a beacon of hope for authentic forms of tourism.

Community-based development is a strategy used by tourism planners to mobilize communities into action to participate in broadening the scope of offerings in the industry. The goal is socio-economic empowerment and a value-added experience for local and foreign visitors. This process opens new niches for destinations, most notably for the nature, culture, and adventure travelers. What this achieves is a policy objective of creating a culture of inclusion in the industry, whereby communities participate and share in the wealth of the industry, dispelling a long-held perception of tourism as an exploiter of wealth where only the rich can benefit.

Community-based development empowers people to be more aware of the value of their community assets-their culture, heritage, cuisine, and lifestyle. It mobilizes them to convert these into income generating projects while offering a more diverse and worthwhile experience to visitors. Every citizen is a potential business partner to be trained in small business management, environmental awareness, product development and marketing. This type of 'people-centered' tourism promotes a sense of 'ownership' which augurs well for the industry's sustainability.

Continuing with the above line of thinking, community-based tourism (CBT) is may be defined as tourism that takes environmental, social, and cultural sustainability into account. It is managed and owned by the community, for the community, with the purpose of enabling visitors to increase their awareness and learn about community and local ways of life. Existing terms like heritage tourism, eco-tourism, agri-tourism, cultural tourism, etc., can all be forms of the community tourism product, within the constraint that these are to be promoted with the spirit of community centeredness and sustenance.

This chapter attempts to weave together the theory and practice of CBT. Though the idea of community-based development is age-old, its adoption into tourism is relatively recent. The chapter examines how CBT can be a panacea for many of the evils of the mainstream industry driven mass-tourism. It exhorts that, while community is at the center stage, there are important roles for governments and non-governmental organizations in CBT. The role of public-private partnership (PPP) is also examined.

11.2 WHY SHOULD WE NEED A COMMUNITY-BASED MODEL?

The key benefits of CBT are seen to be: job creation; poverty reduction; less impact on an area's culture and environment than that exerted by mass tourism; community capacity building and pride; and revenue for maintaining or upgrading the community's cultural assets. The relationship between resources and actions in CBT is illustrated below:

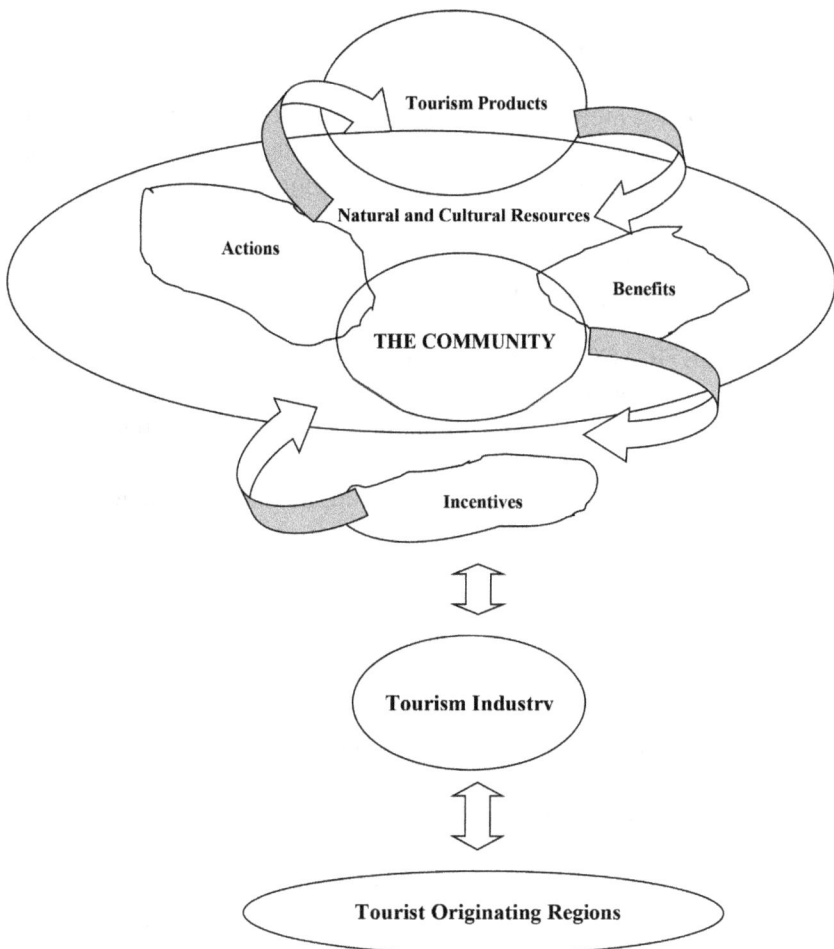

The key rationale underlying the approach and objectives of CBT for conservation and development is that CBT through increased intensities of participation can provide widespread economic and other benefits and decision-making power to communities. These economic benefits act as incentives for participants and the means to conserve the natural and cultural resources on which income generation depends. Note from the diagram that the community is at the center and is occupying the commanding position with regard to the management of its natural and cultural resources which can be reformulated as tourism products. The relationship of the industry to the tourism products developed out of the natural and cultural resources of the community is not a direct, one-to-one relationship; on the contrary, it is through the intermediation of the community. This is aimed to ensure that the aspirations of the community are never bypassed by the extraneous industry interests. One can see CBT as an interaction among the three major groupings of the community, the tourism industry, and the tourists themselves. In the language of cost-benefit analysis (CBA), community-based tourism may be expressed in terms of the following inequalities:

    i.   CB>IB>IC>CC (The inequality of interaction between the community and the industry);

    ii. CB>TB>TC>CC (The inequality of interaction between the community and the tourists);

    iii. IB>TB>TC>IC (The inequality of interaction between the industry and the tourists).

Note: CB: Community benefits; CC: community costs; IB: industrial benefits; IC: industrial costs; TB: tourist benefits; and TC: tourist costs.

Note that the maximum benefit with the least cost goes to the community. This is the essential condition for CBT and the other conditions are not as consequential as this. The third condition may even be contested; however, we feel this is desirable since only if the industry benefit is significant than the industry cost shall it survive and not move on to unsustainable practices; tourists shall visit a CBT destination as long as tourist benefit is more than tourist cost.

11.3   THE BLUEPRINT FOR DEVELOPING COMMUNITY-BASED TOURISM

In order that CBT be developed in a systematic manner, a methodological framework needs to be adopted. An outline of a suggested framework is provided below (Tuffin, 2005):

1. Choose a destination;
2. Complete a feasibility study with the community;
3. Create an action plan;
4. Set up an administrative system;
5. Prepare for operation;
6. Monitor and evaluate.

> **Step 1: Choose a Destination:** Choosing an appropriate destination requires collecting information that leads to an understanding of the community. A detailed study of the village context includes collecting information about the community from organizations working there, government agencies, other communities in the area, and the community members themselves.

> **Step 2: Complete a Feasibility Study:** The community needs to be fully involved in the process of deciding if they want to be involved in a tourism project. The process for building consensus in the community requires that the information and data be studied with the public and private partners and then an action plan be formulated. It is important to be open and honest about the limitations of the community when deciding whether to continue or not. The decision to develop CBT must be agreed upon by all parties. During this process the community will be stimulated to think about the reasons and motivations for developing CBT. They should be able to discuss the issues and visit communities which are already involved in CBT. The community members need to answer questions like:

- Do you want CBT to raise income?
- Do you want CBT to preserve culture?
- Do you want CBT to conserve natural resources?
- Do you want CBT to bring more knowledge and skills into the community?

➢ **Step 3: Create an Action Plan:** If all parties reach a consensus, the planning process can begin. In this stage the community creates an action plan and enters into agreement with external agencies like tour operators. Some of the key issues that need to be considered include: Programs for the tourists; Services that will need to be provided; Development of facilities and infrastructure; Training that will need to be provided; Carrying capacity; and, Tour program and price.

The public partner (association of the community members) will need to formulate a monitoring and evaluation plan that includes the associated indicators and the private partner can begin to draft a marketing plan and strategy.

➢ **Step 4: Set-Up an Administrative System:** Without transparent organization, confusion, suspicion, and conflict can arise in the community. It is crucial that the community sets up a clear administrative system to effectively manage CBT.

The administrative organization will focus on the following: Participation level of community members; Division of roles in operation; Division of benefits; Transparency of management; Measures to control economic and social impacts; Measures to control natural and cultural impacts; and, Cooperation, and communication with public and private partners.

➢ **Step 5: Preparation of Operation:** Before full operation of the tour program can start the community and its partners need to acquire skills and experience in operating CBT. The infrastructure must all be designed and built and the equipment acquired and put in place. At this stage emphasis will be placed on:

- **Training:** Including guiding skills, language learning, food preparation, housekeeping, and simple accounting systems.
- **Preparation of Information:** Involving the educational content of the tour program; the things about themselves that the community members will share with tourists.
- **Infrastructure Design and Construction:** Community lodges, trails, water systems, power systems, toilets, etc.

The community members will need to gain experience in guiding and operating the tour program and distributing benefits. It will be necessary to bring pilot groups of tourists into the community so that the community members can see what works and what does not and so that they can practice their skills and test the administrative systems.

> **Step 6: Monitoring and Evaluation:** Monitoring and evaluation starts once the program is in full operation. It helps to identify problems, impacts, and benefits, as well as to ensure the sustainability of the operation. It examines the extent to which the project is meeting its objectives. It should also result in plans and efforts to compensate for weaknesses, correct problems, adjust systems and improve the program. Monitoring and evaluation is a participatory process. All stakeholders should play a role in gathering the monitoring data, assisting in the analysis, and in actions taken as a result of the final assessment and evaluation.

The aspects monitored include: Environmental impacts; Economic impacts; Cultural impacts; Social impacts; Efficacy of CBT as a development tool, etc.

Information can be gathered from the tourists, the community members and from physical inspections of infrastructure and the environment. Tools used for monitoring can include questionnaires, interviews, focus groups, guest books, photographs, checklists, trend lines, seasonal calendars, and so on.

Often in CBT the locations are remote and subject to national policies regulating access by foreigners as well as domestic visitors. Thus, while initial assessments show considerable potential as tourism destinations, there may be regulations that restrict access by numbers and by seasons. International policies and actions have complex linkages with the visitor to a protected area and the local entrepreneur. Political instability can also affect the volume of visitors.

11.4   THE ROLE OF NON-GOVERNMENTAL ORGANIZATIONS

Nongovernmental organizations, or NGOs, are generally accepted to be organizations which have not been established by governments or agreements among governments. Since the early 1990s there began to be recognition of the importance of NGOs. They are found to have closer ties to on-the-ground realities in developing countries and, perhaps more important, to be able to deliver development aid considerably more cheaply than states or intergovernmental organizations.

Aided by advances in information and communications technology, NGOs have helped to focus attention on the social and environmental externalities of business activity. Multinational brands have been acutely susceptible to pressure from activists and from NGOs eager to challenge a company's labor, environmental or human rights record. Even those businesses that do not specialize in highly visible branded goods are feeling the pressure, as campaigners develop techniques to target downstream customers and shareholders. In response to such pressures, many businesses are abandoning their narrow shareholder theory of value in favor of a broader, stakeholder approach which not only seeks increased share value, but cares about how this increased value is to be attained. Such a stakeholder approach takes into account the effects of business activity-not just on shareholders, but on customers, employees, communities, and other interested groups.

There are many visible manifestations of this shift. One has been the devotion of energy and resources by companies to environmental and social affairs. Companies are taking responsibility for their externalities and reporting on the impact of their activities on a range of stakeholders. Nor are companies merely reporting; many are striving to design new management structures which integrate sustainable development concerns into the decision-making process. NGOs have been involved with tourism related issues for a long time. Current international activities in community tourism planning and development reflect a strong interest and involvement by NGOs, particularly those focused on resource conservation. Traditionally, they have been critical. They have campaigned against contentious issues, such as tourism links to child prostitution and the forced relocation of peoples for new developments. NGOs have frequently been concerned with tourism related environmental issues, opposing, for example, the establishment of golf courses in developing world locations that have exploited land and water previously available to local communities. NGOs have also focused their attention on the flow of income from tourism, particularly in the developing world, examining how this has been generated and how equitable its distribution is. Most NGOs commenting on tourism have had their major interests elsewhere.

However, NGOs can play an important constructive role in the development of management strategies and in the planning process of tourism development for various reasons. Tourism is becoming far too commoditized, and NGO involvement offers alternative ways of viewing the tourism experience. NGOs have prioritized development approaches that include host community perspectives, emphasized host-visitor interaction and stressed nature and cultural conservation. A number of NGOs have been actively involved in tourism related projects. The motives of NGOs in these circumstances appear to have been to boldly go where government or private commercial organizations would find political or economic resistance. Being nongovernmental institutions, they can establish and facilitate the participation of local stakeholders. Being non-profit organizations, they can promote the sustainable use of biodiversity and cultural resources and point out the economic benefits of the integration of tourism development and nature/culture conservation. Being experts in ecological sciences, social development, and project management, and having a deep insight into the regional political and economic structures, NGOs can contribute significantly to the sustainability of community-based tourism development. NGOs can sensitize the public and even organize mass movements if avaricious industrial interests placate the wider goals of community centered tourism development. Thus, they can act as brokers between conflicting groups and some of them have the potential to act in a conflict management capacity.

Another role that NGOs can take over is that of a co-operative agency that manages a community's initiatives towards CBT. Where international tour operators are unable to contract ground services to in-country operations or do not employ residents of that country, the amount retained in the destination is obviously lower than if this was not the case. Some NGOs appear to have special skills in collaborative partnerships based upon shared aims with local communities, the private sector, and other NGOs. The NGO can become a unified marketing front for the small and medium scale tourism enterprises run by the community members. With the help of e-commerce technologies, such a body can bypass the middlemen like travel agents and tour operators and reach the tourist originating markets directly. With tour operators invariably demanding payment in the visitor's country of origin, the benefits to national economies can vary considerably. This has the potential to make the community's tourism offerings more cost-competitive, too. In these situations, the development work of NGOs can be directly supported by tourism income flows, active tourist participation in projects, or through direct and indirect donations to their work.

Thus, NGOs are shown often to be both campaigning and proactive bodies, capable of operating in a wide variety of natural, economic, and political environments. In sum, NGOs in CBT does one or more of the following things:

- Contribute to the development of policies and plans for the CBT industry.
- Assist the government in developing a standard for responsible community-based tourism.
- Assist the government, private sector, and communities in implementing, monitoring, and evaluating CBT.
- Attract funding from donor agencies to develop specific CBT projects.
- Assist communities and community groups in organizing themselves, preparing themselves for CBT and implementing CBT projects.
- Assist the government in conducting tourism and environmental awareness programs among communities and the tourism industry at large.
- Liaise between the private sector and communities to generate more community involvement in the tourism sector and stronger private sector commitment.
- Deliver education, training, bridging courses, and other capacity building exercises to local communities.
- Resist against inequitable tourism development by campaigning and mobilizing community support.
- Manage and market the community tourism product for the community, at least until the community gains experience to manage on its own.

Some of NGOs that do outstanding work in the area of tourism are:

- Tourism Concern (www.tourismconcern.org.uk);
- Equations in India (www.equitabletourism.org);
- Ecumenical Coalition on Third World Tourism (www.ecotonline. org);
- World Wide Fund for Nature (www.wwf.org);
- Conservation International (www.conservation.org);

- Rainforest Alliance (www.rain-forest-alliance.org);
- Cultural Restoration Tourism Project (http://home.earthlink. net/~crtp);
- International Council on Monuments and Sites (www.icomos. org);
- International Society for Eco-tourism Management (www. ecomanage.com);
- Partners in Responsible Tourism (www2.pirt.org);
- Retour Foundation (www.retour.net).

11.5   GOVERNMENTS AND COMMUNITY-BASED TOURISM (CBT)

Community Based Tourism cannot be implemented successfully without the constant and coordinated facilitation by the various governmental bodies. Governments play a critical role through their institutional leadership, guaranteeing stakeholders' participation. The Governments' role is also essential in the establishment of regulatory and policy frameworks, ensuring their enforcement, the application of appropriate economic instruments (including the removal of environmentally perverse subsidies), and monitoring environmental quality. What communities do in tourism depends on the opportunities and power they have, the incentives and prices they face, and their access to skills, training, capital, and markets. All of these are shaped by government policies, regulation, and taxes. Only governments can provide the strategic planning base for CBT which is so clearly needed. Only they can ensure that valuable and fragile habitats are identified, that baseline studies and monitoring are carried out, and that overall infrastructure needs and implications are assessed. And only they can establish emissions standards and citing and design requirements, and ensure that they are enforced.

Governments need to make resources such as national tourist boards accessible to CBT operations, while ensuring that systems of licensing or tourism standards do not act as barriers. Government, especially, the local government, provides the core utilities and infrastructure on which the tourism industry is based. This includes district and city roads, lighting, water, and sewerage, public transport systems, signs, airports, and ports. If local government operates attractions such as museums, art galleries, sports stadiums, convention centers, parks, gardens, events, tours, and other amenities, the same become additional motivators for tourists to visit a CBT destination. The government can integrate travel information about the CBT destinations in the country into its Visitor Information Network, too.

Private companies cannot be expected to share profits and power with rural communities simply because it is a kind thing to do. But governments can create the conditions under which it is in their interests to work with communities by giving communities market power and giving the private sector more security of investment and incentives for partnership. This can be by means of:

- asking private sector bidders to develop proposals for community partnership, and making these key criteria in allocating tourism rights. This small change to the planning process can force every new investment to address community tourism issues;
- devolving tenure to communities, to give them market power in forming agreements;
- giving communities an equity shares in government-private agreements;
- helping local residents to start private enterprises.

Policies vary from country to country and over time. It is often the overall approach that is most important in helping community tourism to flourish. Some tips that are suggested from governments in CBT by Africa Resources Trust (ART, 2005) are:

- Create supportive attitudes in government;
- Let communities develop tourism over time;
- Create opportunities and remove constraints, rather than plan community tourism for them;
- Recognize that local people will have multiple livelihood objectives, not just maximizing cash income. Concerns about how land or natural resources are used, or access to training, can be equally important to livelihoods;
- Enhance their power in the tourism market;
- Ensure tourism sector regulations encourage rather than exclude the informal sector;
- Welcome NGO facilitation-it is usually needed.

There are two extremes to be avoided: one is to ignore community tourism or pretend it will happen with no support from Government and the industry. The other is for government and the industry to try to do everything and do it now, without allowing time for local people to develop their ideas and skills.

11.6 PUBLIC-PRIVATE PARTNERSHIP (PPP) FOR COMMUNITY-BASED TOURISM (CBT)

Public-private partnership (PPP or P3) is a variation of privatization in which elements of a service previously run solely by the public sector are provided through a partnership between the government and one or more private sector companies. Unlike a full privatization scheme, in which the new venture is expected to function like any other private business, the government continues to participate in some way.

Important variants of PPPs are (Source: National Council for Public Private Partnerships, USA): build/operate/transfer (BOT) or build/transfer/operate (BTO); build-own-operate (BOO); buy-build-operate (BBO); service contracts (SC); design-build (DB); design-build-Maintain (DBM); design-build-Operate (DBO); developer finance: enhanced use leasing (EUL); lease/develop/operate (LDO) or build/develop/operate (BDO); lease/purchase; sale/leaseback; tax-exempt lease; and turnkey arrangement.

1. **Build/Operate/Transfer (BOT) or Build/Transfer/Operate (BTO):** The private partner builds a facility to the specifications agreed to by the public agency, operates the facility for a specified time period under a contract or franchise agreement with the agency, and then transfers the facility to the agency at the end of the specified period of time. In most cases, the private partner will also provide some, or all, of the financing for the facility, so the length of the contract or franchise must be sufficient to enable the private partner to realize a reasonable return on its investment through user charges.

   At the end of the franchise period, the public partner can assume operating responsibility for the facility, contract the operations to the original franchise holder, or award a new contract or franchise to a new private partner. The BTO model is similar to the BOT model except that the transfer to the public owner takes place at the time that construction is completed, rather than at the end of the franchise period.

2. **Build-Own-Operate (BOO):** The contractor constructs and operates a facility without transferring ownership to the public sector. Legal title to the facility remains in the private sector, and there is no obligation for the public sector to purchase the facility or take title. A BOO transaction may qualify for tax-exempt status as a service contract if all Internal Revenue Code requirements are satisfied.

3. **Buy-Build-Operate (BBO):** A BBO is a form of asset sale that includes a rehabilitation or expansion of an existing facility. The government sells the asset to the private sector entity, which then makes the improvements necessary to operate the facility in a profitable manner.

4. **Service Contracts (SC):** A public partner (federal, state, or local government agency or authority) contracts with a private partner to provide and/or maintain a specific service. Under the private operation and maintenance option, the public partner retains ownership and overall management of the public facility or system. Another way is public partner (federal, state, or local government agency or authority) contracts with a private partner to operate, maintain, and manage a facility or system proving a service. Under this contract option, the public partner retains ownership of the public facility or system, but the private party may invest its own capital in the facility or system. Any private investment is carefully calculated in relation to its contributions to operational efficiencies and savings over the term of the contract. Generally, the longer the contract term, the greater the opportunity for increased private investment because there is more time available in which to recoup any investment and earn a reasonable return. Many local governments use this contractual partnership to provide wastewater treatment services.

5. **Design-Build (DB):** A DB is when the private partner provides both design and construction of a project to the public agency. This type of partnership can reduce time, save money, provide stronger guarantees, and allocate additional project risk to the private sector. It also reduces conflict by having a single entity responsible to the

public owner for the design and construction. The public sector partner owns the assets and has the responsibility for the operation and maintenance.

6. **Design-Build-Maintain (DBM):** A DBM is similar to a DB except the maintenance of the facility for some period of time becomes the responsibility of the private sector partner. The benefits are similar to the DB with maintenance risk being allocated to the private sector partner and the guarantee expanded to include maintenance. The public sector partner owns and operates the assets.

7. **Design-Build-Operate (DBO):** A single contract is awarded for the design, construction, and operation of a capital improvement. Title to the facility remains with the public sector unless the project is a design/build/operate/transfer or design/build/own/operate project. The DBO method of contracting is contrary to the separated and sequential approach ordinarily used in the United States by both the public and private sectors. This method involves one contract for design with an architect or engineer, followed by a different contract with a builder for project construction, followed by the owner's taking over the project and operating it.

   A simple DB approach creates a single point of responsibility for design and construction and can speed project completion by facilitating the overlap of the design and construction phases of the project. On a public project, the operations phase is normally handled by the public sector under a separate operations and maintenance agreement. Combining all three passes into a DBO approach maintains the continuity of private sector involvement and can facilitate private-sector financing of public projects supported by user fees generated during the operations phase.

8. **Developer Finance:** The private party finances the construction or expansion of a public facility in exchange for the right to build residential housing, commercial stores, and/or industrial facilities at the site. The private developer contributes capital and may operate the facility under the oversight of the government. The developer gains the right to use the facility and may receive future income from user fees.

   While developers may in rare cases build a facility, more typically they are charged a fee or required to purchase capacity in an existing facility. This payment is used to expand or upgrade the facility. Developer financing arrangements are often called capacity credits, impact fees, or extractions. Developer financing may be voluntary or involuntary depending on the specific local circumstances.

9. **Enhanced Use Leasing (EUL):** An EUL is an asset management program in the Department of Veterans Affairs (VA) that can include a variety of different leasing arrangements (e.g., LDO, build/develop/operate). EULs enable the VA to long-term lease VA-controlled property to the private sector or other public entities for non-VA uses in return for receiving fair consideration (monetary or in-kind) that enhances VA's mission or programs.

10. **Lease/Develop/Operate (LDO) or Build/Develop/Operate (BDO):** Under these partnerships' arrangements, the private party leases or buys an existing facility from a public agency; invests its own capital to renovate, modernize, and/or expand the facility; and then operates it under a contract with the public agency. A number of different types of municipal transit facilities have been leased and developed under LDO and BDO arrangements.

11. **Lease/Purchase:** A lease/purchase is an installment-purchase contract. Under this model, the private sector finances and builds a new facility, which it then leases to a public agency. The public agency makes scheduled lease payments to the private party. The public agency accrues equity in the facility with each payment. At the end of the lease term, the public agency owns the facility or purchases it at the cost of any remaining unpaid balance in the lease.

    Under this arrangement, the facility may be operated by either the public agency or the private developer during the term of the lease. Lease/purchase arrangements have been used by the General Services Administration for building federal office buildings and by a number of states to build prisons and other correctional facilities.

12. **Sale/Leaseback:** This is a financial arrangement in which the owner of a facility sells it to another entity, and subsequently leases it back from the new owner. Both public and private entities may enter into a sale/leaseback arrangement for a variety of reasons. An innovative application of the sale/leaseback technique is the sale of a public facility to a public or private holding company for the

purposes of limiting governmental liability under certain statues. Under this arrangement, the government that sold the facility leases it back and continues to operate it.

13. **Tax-Exempt Lease:** A public partner finances capital assets or facilities by borrowing funds from a private investor or financial institution. The private partner generally acquires title to the asset, but then transfers it to the public partner either at the beginning or end of the lease term. The portion of the lease payment used to pay interest on the capital investment is tax exempt under state and federal laws. Tax-exempt leases have been used to finance a wide variety of capital assets, ranging from computers to telecommunication systems and municipal vehicle fleets.

14. **Turnkey Arrangement:** A public agency contracts with a private investor/vendor to design and build a complete facility in accordance with specified performance standards and criteria agreed to between the agency and the vendor. The private developer commits to build the facility for a fixed price and absorbs the construction risk of meeting that price commitment. Generally, in a turnkey transaction, the private partners use fast-track construction techniques (such as DB) and are not bound by traditional public sector procurement regulations. This combination often enables the private partner to complete the facility in significantly less time and for less cost than could be accomplished under traditional construction techniques.

In a turnkey transaction, financing, and ownership of the facility can rest with either the public or private partner. For example, the public agency might provide the financing, with the attendant costs and risks. Alternatively, the private party might provide the financing capital, generally in exchange for a long-term contract to operate the facility.

Further details about these variants are beyond the scope of this chapter. However, readers desirous of knowing the same are advised to consult the website of the National Council for Public Private Partnerships, USA (http://ncppp.org).

There are several basic characteristics of CBT development that make PPPs a possibility (UNESCAP, 2001). First, communities may not have the skills and experience in tourism management. Second, community tourism ventures take time to set up and require a process of intensive capacity building. Third, community tourism ventures may not be profitable when they are initiated.

Partnership is becoming a powerful tool for implementing CBT policies more effectively. PPPs enable the public sector to benefit from commercial dynamism, the ability to raise finances in an environment of budgetary restrictions, innovation, and efficiencies, harnessed through the introduction of private sector investors who contribute their own capital, skills, and experience. The positive characteristics of PPP arrangements for infrastructure development appear particularly attractive to developing countries like India given the enormous financing requirements, the equally large funding shortfall, the need for efficient public services, availability of a pool of private finance, growing market stability and privatization trends creating a favorable environment for private sector participation (Subramaniam, 2005). Despite numerous advantages, certain negative aspects—a too large role for governments, partnerships lacking attention to market needs, disproportional investments, inefficiency of public administration, institutionalization of projects and lack of creativity—have to be taken into consideration before going in for any PPP based tourism development since these are antithetical to the spirit of CBT.

According to OECD (1997), good partnership involves a clear definition of roles, competencies, responsibilities, and advantages both in public administrations and private enterprises. In particular, the public sector, as an agent of development, may help achieve optimal exploitation of public resources and services, safeguard the environment, and develop human resources. Partnerships must be based on agreements which show the economic benefits for the public/private sector and/or center/periphery. More and more forms of partnerships are developed in almost all areas of tourism policy. Governments have to play an important role in new ways of organizing this co-operation, notably by defining a clear national strategy for tourism policy which will stimulate and guide innovative partnerships and give incentives to all individual partners to participate in the development.

11.7 CONCLUSION

Raising some awareness of stakeholders means that CBT is essential for promoting a clear understanding of tourism sustainability. Of course, tourism sustainability associated to awareness rising allows locals higher levels of self-determination and autonomy respecting other global forces. Having said this, awareness campaign seems to be equally important to the involved stakeholders. Over years, development professionals attempted to encourage CBT since the 70s decade. During the eco-tourism book, in the middle of the 90s decade, CBT occupied a central position in the literature as well as the interests of experts. Nonetheless of this fact, few projects have impacted positively in community. This happens simply because CBT is complex and dynamic, situating as a green area of study. Continued information sharing of research is at least requested to find new solutions to the problems of the industry. Continuing research through the provision of financial and technical supports is vital to understand the positive effects of CBT in the challenging times the industry today faces.

KEYWORDS

- **build/transfer/operate**
- **build-own-operate**
- **community based tourism**
- **inclusive tourism**
- **sustainability**
- **synergy**
- **tourism**

REFERENCES

Africa Resources Trust, (2005). *Community Tourism in Southern Africa.* Accessed from https://resourceafrica.net/about-resource-africa/ (accessed on 18 October 2021).

Brandon, K., (1996). *Ecotourism and Conservation: A Review of Key Issues.* World Bank Environment Department. Paper No. 033. Washington. DC: World Bank.

Cooperrider, D. L., (1990). Positive image. Positive action: The affirmative basis for organizing. In: Srivastava, S., Cooperrider, D. L., & Associates, (eds.), *Appreciative Management and Leadership.* San Francisco: Jossey-Bass.

Liburd, J. J., (2004). NGOs in tourism and preservation - democratic accountability and Sustainability in question. *Tourism Recreation Research, 29*(2), 105–110.

OECD, (1997). *OECD Conference on Partnerships in Tourism.* Conference organized by the OECD, the Italian Department of Tourism and the City of Rome.

Pretty, J. N., Gujit, I., Scoones, I., & Thompson, J., (1995). *A Trainer's Guide for Participatory Learning and Action. Sustainable Agriculture Program.* International Institute for Environment and Development, 3 Endsleigh Street. London WCIH ODD, UK.

Responsible Ecological Social Tours, (2005). Accessed from www.rest.org.th (accessed on 1 October 2021).

Subramaniam, P., (2005). Partnership in tourism. In India revisited: A symposium on reorienting our policy on tourism. *Seminar*, 19–24.

The Mountain Institute, (2000). *Community-Based Tourism for Conservation and Development: A Resource Kit.* Washington: The Mountain Institute.

TPDCO, (2005). *Tourism Product Development Company: What is a Community-Based Tourism Project?* Accessed from: www.tpdco.org (accessed on 1 October 2021).

Tuffin, B., (2005). *Community-Based Tourism in the Lao PDR: An Overview.* Accessed from www.nafri.org.la (accessed on 1 October 2021).

UNESCAP, (2001). *ESCAP Tourism Review No. 22: Managing Sustainable Tourism Development.* Accessed from: www.unescap.org (accessed on 1 October 2021).

Wells, M., Brandon, K., & Hannah, L., (1992). *People and Parks: Linking Protected Area Management with Local Communities.* World Bank. WWF. USAID. Washington D.C.

Whelan, T., (1991). *Nature Tourism: Managing for the Environment.* Washington, D.C.: Island Press.

Worah, S., Svendsen, D., & Ongleo, C., (1999). *Integrated Conservation and Development: A Trainer's Manual.* WWF-UK, Godalming.

CHAPTER 12

Could Information and Communication Technologies Be the Hope for Third World Tourism?

ABSTRACT

As discussed in earlier chapters, the future of tourism remains uncertain in the short-run. The COVID-19 has accelerated a long-dormant crisis in the tourism industry. The fact is experts and policy makers call the attention on the adoption of new opportunities, tactics, and strategies. Digital technologies and ICT (Information communication tech) offer fertile grounds to optimize destinations and consumptions in a paralyzed world. The most important element of the developmental debates in the last decade among social scientists, policy-makers, civil sphere activists has been the growing world-wide inequalities especially in the context of the evolving networked society or the "informationalization of economy" (Hall, 1996). With the rise of the information society, these inequalities are becoming more and more visible. Despite of the widespread disagreement about the concept, the social-economic-political changes generated by the new information technologies are pervasive. Globalization only with the appearance of the information society become completed. Dealing with the emergence of the new social organization and integration forms, it is necessary to consider the impact of information technology generated development in a holistic manner without losing the rich context in which the changes are being taken place.

12.1 INTRODUCTION

The benefits of a new business paradigm based on internet commerce have extensively become a topic of spirited discourses, almost got embedded into the common consciousness of the business communities of many industries, but, deplorably, the same cannot confidently be said of the community of leisure and tourism professionals. A lack of body of knowledge and a higher-level frame to organize issues about information and communication technology (ICT) applications and impacts in tourism, both positive and negative, many of which already have started appearing, is deplorable given the magnitude of consequences of such issues upon tourism which is one of the most information intensive (Werthner and Klein, 1999) and the single largest e-commerce industry in the world. The chemistry of innovative ICT applications with the organizational, social, political, and economic conditions that are likely to support their effective use or misuse and issues of allied importance have started appearing in social science and policy literature, but no study worth to name has appeared that take into account the uniquely rich contexts of specific industries like tourism. Towards this end, the present chapter focuses on ICT applications in tourism that are expected to assist developing countries to reap the socio-cultural, environmental, and economic benefits associated with the extremely rapid innovations being taken place in advanced ICT and attempts to plot a blueprint of mechanisms that bring together policy, supplier, user, and third-party communities that are essential in ensuring that the promises are met.

The Internet substantially alters the patterns of the tourism sector and potentially brings all participants to the market. The marked change in the present era from the initial days of seriously employing ICT, i.e., started with computerized reservation systems (CRS), is that the fear that a few CRS/GDS owners, some of them airliners, would overpower all the rest, is no longer panicking us. More and more destination marketing organizations (DMG) are promoting their destinations online and take the Internet as one part of their marketing strategies (Buhalis, 2000). Internet has enough room not only for big travel portals like Expedia and Travelocity but also for the websites of small-scale hospitality service providers located somewhere at the remotest interiors of Africa too. Tourism industry has enthusiastically embraced the Internet and used it as a marketing and information communication channel (Yuan and Fesenmaier, 2000). Specialized segments of tourism, organized around small-scale family run units, have proliferated in the 1990s and later, and Web technologies provide an unexpected opportunity to market the services globally. This kind of tourism, if it can be developed and managed properly, is an appropriate way to generate income from the cultural or natural features of a country. The revenues generated by tourism of this sort are captured by those service providers who are stewards of those assets and are used in part to maintain and manage the assets. And hence, Internet is effectively synthesizing the so far dichotomous and irreconcilable discourses of economic and environmental sustainability.

12.2   THE INFORMATION SOCIETY: MOVING ONE STEP CLOSER TO NATURE

Information as the key determinant of success in life has increasingly been true in our times than any time before and our society is uniquely characterized by the widespread adoption and use of ICT into all possible realms of not only public but also private life. As the Internet becomes firmly implanted in the public's mind, people increasingly expect to find the information they are looking for, and buy the goods they want, online. Organizations in the public domain, whether commercial or otherwise, risk losing credibility if they do not enable people to do this.

This change is not the accidental boon of a good day morning. It began with attempts to translate the component tasks of many of man's traditionally held jobs into mechanical forms and structures. Internal feedbacks as well as environmental controls were instituted and such a design was able to collect information from various sources and adjust the production process by itself or through consoles in the system. Number crunching machines of an earlier generation could easily substitute the operator w.r.t control by feedback; but the nature of data from the external environment, its collection, processing, and incorporation remained extremely difficult to be routinized so much so that the static mathematical models of management science (only which the computing machines of early generations could handle) continued to remain ineffective when it came to the need to integrate business decisions with concerns of wider stakeholders and the macro-environment in general. Convergence of computation and communication technologies is the point of radical departure here with which the dictum that 'everything is linked with everything else' that was appropriate hitherto only in the matter of natural environment started becoming true in the case of socio-technical environment too. That is, the new man-made environment reconstituted in terms of socio-technical system was made one step closer to its natural counterpart. Hitherto society and social organizations, especially western conceptions of them, were a rebellious logic to nature but the possibilities offered by ICT (of course in conjunction with other technologies, say, biotechnology) make them a 'natural' answer. Exploitation of nature and natural resources is never an item in the intrinsic agenda of humankind, but it just reflects the cultures, ideologies, and evolutionary constraints. According to A. Toffler, meaning of the natural environment is becoming more important as the world seeks new values resulting in changes of cultures and ideologies associated with the rapid progression of what he termed as the 'third wave' at the coffin of the withering industrial civilization. The important question for tourism academics and practitioners here is, will tourism industry be a sustainable innovator in the new wave by internalizing the spirit of the new times or not. Enough preliminary evidence is available to show the shifting orientation of tourism from mass tourism to more sustainable forms that betray the rhyme and rhythm of the new wave. The main causal agent and enabler to this qualitative shift is identified by many as information and communication technologies (ICT).

Knowledge is 'Power' when a few have asymmetric access to it. By allowing access to all knowledge of all kinds, the Internet is redefining knowledge as 'justice.' The potential of ICTs in realizing a sustainable world order around justice, equity, and probably grace is noted by many futurists. And tourism, due to it is being an information intensive industry, is one area where the first implications of all these can appear. At the points of sale, experience, and recall, tourism is little more than an information product.

12.3   SUSTAINABLE TOURISM AND ITS OBJECTIVES

The objective of sustainable tourism is to ensure both intra and inter-generational equity that may melt down into the operational goals like this:

- Promoting awareness of how people's actions affect the environment and future recreation use;
- Enhanced economic well-being and quality of life for host communities;
- Involving host communities in tourism projects and initiatives;

- Engendering a sense of local pride and ownership in order to modify behavior in appropriate ways;
- Increased provision and uptake of public transport by visitors and its efficient management.

Dominant sustainable tourism debates seek to change the nature of economic growth rather than limit it, for the contributors of which the challenge of sustainable development is to find new products, processes, and technologies which are environmentally friendly while they deliver the things we want. At the heart of the debate over the potential effectiveness of technologies for sustainable development is the question of whether technological change can reduce the negative impact of tourism development from the level that would have happened from the same degree of development and at times whether tourism development itself can be re-conceptualized in a qualitatively better manner. Such questions are then linked to more radical debates to posit that such an accomplishment would require more than just a few adjustments to existing technological systems; it would require a radically different technology; but since technology is socially shaped, as modern scholars of technology studies argue, radical technological change cannot take place without equally radical social change. So, it is important not to put too much emphasis on technological factors without considering the social, political, and economic factors that can be crucial in the shaping and implementation of technologies.

12.4 MARKETING SUSTAINABILITY FOR SUSTAINABLE MARKETING: WHAT GOES WRONG?

Marketing, according to most revered gurus like Kotler, starts with the customer: Understand his mental model, design, and develop products and services befitting the same, arrange channels of distribution and communicate the same is all about the game of marketing. Tourism, in spite of its unique dimensions, has received the same treatment so far but for a few exceptions. Modifying a tourism product, especially if it is an environmental good or a cultural object to suit the whimsical demand of the market is both unethical and rampageous.

In the case of tourism products, the most authentic product is the row, unmodified one. However, marketers have ever been successful in manipulating the media for the message causing blurred boundaries between reality and imagination in the minds of customers, the scope of which is multi-fold now with the arrival of the 'new' media. And most of the research already done betrays this trend evidently (see Hoffman and Novak, 1996; Ho, 1997; Hoger, Cappel, and Myerscough, 1998; Sterne, 1999; Kaye and Medoff, 1999; Hofacker, 2000). The ICT revolution and the paradigmatically altered world-view in the post-Einsteinian era of ours has impacted in such a way that people have lost faith in meta-narratives and no frames of reference is any longer accepted as authentic and derived from a platonic ideal elemental frame. Hidden herewith is also an opportunity: Adjust the fourth P (Promotional tools) instead of altering the other Ps, especially the Product aspect, for the immediate gain and deliberately design the message so as to permit multiple read-ings. So, it is to be accepted that the media construction of reality if the same were from the point of view of the destination communities and not infected with the profit-oriented designs of intermediaries in the supply chain and other external agents, merits, at least in terms of an 'instru-mental rationality' to be listed as a better representation of the real and authentic. Also, from the angle of sustainable development of tourism, the concern is whether the new technological artifacts are better posited and probably qualitatively different in aiding to attain the local need for sustainable livelihood through promoting e-tourism while fostering socio-cultural and natural capital which is the very raison d'etre that makes the destination a destination.

The attainment of these objectives depends largely on being able to involve, inform, and educate visitors and local people in order to modify behavior in appropriate ways. Visitors must be informed and aware of the impact of their actions throughout the visit process, from initial choice of destination and mode of transport to choose of on-site activity and accommodation. Local people need to be fully aware of the impacts of tourism (both positive and negative), and the importance of looking after the natural and cultural heritage of their area, which is the fundamental tourism resource. More importantly, the decision as to what to 'educate' them should emerge from the very same feedings from these groups. Local Tourism professionals, like their colleagues in other pedantic communi-ties, have so far tried, but in vain, to construct 'etic' scales of sustainability in absolutist foundations and have totally neglected many of these voices while demanding all of them to obey the inviolable commandments for sustainability.

So, one of the keys means to look for an answer to the question of the role of the Net in promoting sustainable tourism is to examine how using e-commerce stakeholder groups for whom sustainability is 'really' meaningful can feed that meaning, probably multiple meanings, as guidelines and perspectives into the sustainability criteria of the policy making bodies. Another, which is more visible, is the role of the new media in disseminating the conservation agenda to the tourists.

The emergence of new forms of social relations among the community of travelers via the technology of the Internet and its embeddedness in the social relations of production creates a new dynamic. The access to the technology of the Internet which is a component part of the means of production, is a historically significant development. This combined with a rising awareness of our global interconnectivity may represent a potentiality for a new form of touristic consciousness to emerge. This global consciousness connected by the Internet could exert considerable power against repressive exploitative systems of middlemen that thrived over the exceedingly high ownership cost of production of the old capitalism. The ability of this group to act in connected collective action is clearly available by virtue of the Internet. The potential, the means and the intelligence are in place. "Voice of population is the voice God" has remained an unattainable maxim even while the policy-making system seemed flexible enough to accommodate multiple tongues, thanks to the intermediary barriers that 'noisified' any such attempts. Gone are those centuries! The revolutionary strides in technology have been redefining the scope of our rational expectations and it is becoming possible than ever before to make your voice heard and presence felt into the kernels of labyrinths that determine our common future. Local communities in the west have successfully used the power of the Internet in ensuring that they are heeded in all matters that affect them. Now, they can search for practices around the world, successful, and sustainable; seek expert assistance in developing a 'glocal' developmental model; view published policy documents and management plans; communicate dos and don'ts or codes of conduct to the market forces; and so on and so forth.

12.5 COMMUNICATING EXPECTATIONS

Use of the Internet as a tool for promoting sustainable tourism is bound to change once people are more fully aware of its benefits and capabilities.

The Traditional methods of informing and educating visitors, such as leaflets, posters, visitor centers and outdoor panels are effective over a relatively limited geographical area and timeframe. By contrast, the Internet is able to reach a potentially global audience over an indefinitely extended timeframe. Communicating important management or heritage interpretation messages to visitors is an essential part of modifying behavior in order to lessen environmental impacts and generate support for conservation work. Virtual tours and virtual reality reconstructions are one of the greatest potentials that the new media offers. VR tours may be employed as a means to prepare the visitor to an archeological site, a sensitive rural community, or an eco-fragile landscape. Also, it may reconstruct the lost past and feast the discerning traveler. This can serve as a supplement to educational and FAM tours too.

Molding and Reaching Eco-conscious Consumers through dedicated portals has been on the increase during the last few years. Many Green Tourism Associations have taken birth with this purpose and some are really fulfilling the promise. Their promotional as well as educational pieces inform about many aspects of the destination that are not normally included in the main stream tourism literature, therefore increasing quality visitation to these destinations/businesses.

Neoliberal economic theory suggests global market integration as a strategy to reduce poverty. As a development option imposed by the transnational tourist industry, tourism as a world system creates new centers (i.e., the former periphery) while simultaneously creating new peripheries. In a finite world with a limited hinterland for such a continuous expansion, this cannot be sustainable and hence the need for virtual spaces as an alternatives-as playthings for the conspicuous consumption needs of the tourist-carnivore is also highlighted.

12.6 A MORE SUSTAINABLE, ECO-CONSCIOUS TRANSPORT LOGISTICS FOR TOURISM

All humans are inherently mobile. Humans are not tied to a particular place but are able to move if they so desire from one place to another. Mobility broadens our horizons and enables us to establish new contacts, to gain new experiences, and to increase our knowledge. It can help us to cross borders and to extend fellowship. Throughout the world, in the North as well as in the South, motorized mobility is continuously increasing. More and more passengers and goods are being transported over ever-growing distances. Industrialized nations are the driving force of this development. Motorized transport—in particular the automobile-has become an integral part of almost all societies in the world. But we cannot continue to turn a blind eye to the issue of mobility. The laissez-faire attitude that condoned the development of the last decades cannot be upheld any longer. The damage caused by mobility is becoming more and more obvious. The "mobilization" of humankind is taking its toll on present and future generations. An end to this development is, however, not a far cry. E-business will allow for a more collaborative and integrated approach by logistics operators. Intelligent e-agents would allow for the tracking of logistics functions to ascertain effectiveness and efficiency. Intelligent transport systems (ITS) will allow the automated routing and rerouting of carriers minimizing congestion, pollution, cost, and time while improving road safety considerations. Advanced transport telematics (ATT) systems appropriate for all transport modes and their interconnections, including road, air, rail, and water, are being developed to create an integrated transportation infrastructure and once developed to a critical level, these have the potential to redefine humanities whole idea of mobility itself. It is contended that electronic technologies will lead to relocation of jobs to 'clean and green' environments and now there virtually need no more to have a parcel van for the transportation of information goods. These applications can help to increase the competitiveness of rural and remote areas and tele-services in future may provide opportunities for tele-shopping, on-line reservation services, entertainment, and commercial information. Ready and adequate access to information, knowledge, and telecommunications in rural areas would discourage and even arrest urbanization.

In their study of sectoral and regional trends, Dutta et al. (1998) demonstrate that European businesses may transform into virtual enterprises where individual employees perform their work with customers, with suppliers, and with each other in a variety of locations, such as their home, car, office, or public transportation. The same holds true for American enterprises, for which Boudreau et al. (1998) indicate that "going global" means the effective use of information technology. It is time we urgently need to rethink mobility in terms of sustainability for social, economic, and ecological reasons. In order to guarantee sustainability, we must reconsider transportation.

12.7   TRAVEL TELEMEDICINE

When one picture oneself treading ancient highways, climbing up the Himalayas, voyaging through the deep seas or watching the sun sparkle on distant seas…no one ever pictures themselves becoming ill during the journey flitting between bathrooms and spending days feeling too sick to leave their hotel rooms. Still others will have the misfortune to encounter more rare, but potentially deadly, tropical diseases. Adventure tourism is not even in the 100th hypothetical list of a major population of tourists due to the fear of sorts and telemedicine can come as a major boost in invigorating the demand for tourism to areas of potential threat for the traveler. Since the marginal costs for such systems installed for the tourists at remote places are less, the same, probably through the intervention of public sector policy making bodies, can cater to the medicinal requirements of the local population too.

12.8   LOCAL FIRMS, GLOBAL REACH

Literature too often cites the leakage effects of the tourist spending. The power wooed by the intermediaries in the distribution chain of the tourism product and the multinational hospitality firms at the destination, mostly homed in the developed countries of the Western Europe and the US, made the destination with its local and domestic firms, subservient to the wider designs of the former. Whatever a foreign tourist spends could leak out when the package is designed not to involve the local counterparts and even in those cases where the locals were involved, the multinational boss could unilaterally decide terms of the trade.

Internet era could herald a point of departure- and indeed does. It empowers the citizens of the South with the weapon of information. Now, even very small-scale business start-ups are launched as global firms. Anybody can access the firm from any corner of the world if the firm has presence in the cyber space that costs the firm a 100 dollars or less. Destinations in the third world countries now market themselves sans travel agents causing pull factors in the 'new' tourist in the developed countries who dictate the agent in explicit terms his preferences and priorities leaving no room for middlemen in imposing 'some' package. Automation and enhanced interactivity of the reservation system has increased the share of FITs who bypass the redundancy completely too. And, while the tourist gets a better feel of the authentic, the destination retains its benefits.

12.9   PUBLIC ADMINISTRATION OF TOURISM, PROMOTION OF SMES, AND REGIONAL DEVELOPMENT

Public administration automation is a benefit which should never be under-emphasized. The electronic availability of public information can be of major assistance to small and medium-sized enterprises (SMEs) in tourism in administrative procedures for handling trans-border inbound and outbound travel, tax filings, and business opportunities. So-called 'one-stop' government service kiosks can further increase efficiency of service for SMEs. Cooperation between government and the private sector to implement ICTs to provide citizen access to government information and services, if materialized, stimulates interaction between business and government for the good of both. By creating suitable contents on cyberspace and making it available at info kiosks in their close proximity, preferably in the local language and covering local issues among others, will empower citizens with the knowledge to act to bring about sustainable development. Experience so far suggests that ICTs alone are insufficient for substantial local benefits to emerge and that additional factors such as sound development planning and properly mobilized communities contribute significantly to the success of overall development of the region in a sustainable way. Community based pro-poor e-tourism offers an application specific opportunity for development oriented multi-purpose community tele-centers in regions with tourism potential.

The comprehensive collection and analysis of seasonal and diurnal population demands and of physical and environmental factors in management information systems are useful in establishing development priorities for different period of time which is all the more important for destinations for whom tourism is a supplementary means of income, say for a season when agricultural activates are over. Land information system prepared using geographic information systems (GIS) and remote sensing can help farmers plan their activity and facilitate decision making and planning at the local level ICT applications can strengthen regional cooperation and support information sharing among firms, cooperatives, and SME destination networks once interoperability of various information networks is achieved, thereby increasing the competitiveness of rural and remote areas. Public sector should take the first lead in this and systemize the same.

12.10   CONCLUSION: DEVELOPMENT AS SELF-CULTIVATED AND NOT AS IMPORTED

The implicit assumption of the 'important' role of computerization in promoting economic growth and development needs to be investigated further (Avgerou, 1997). The author has strongly argued elsewhere that it is more important to invest in the cultivation of the patterns of behavior or structural changes that underpin the various technological innovations of modernization than it is to invest in the pervasive uptake of ICTs (George, 2002). Achieving a proper tacit structure at the background suitable to the local nuances of the socio-technical system is unavoidable for any sustainable innovation to take shape. This may seem at the outset as a much easier option than buying the hardware but not really so. As Kunkel (1970) says, "… the major problem of economic development is …the change of those selected aspects of man's social environment relevant to the learning of new behavior." Modifications at the tacit level is made further complicated because knowledge transfer at this level is not at all possible via the linear employment of a formal language-needless to say its translation into wisdom and insights for informed action!

The ideal objective of a Zero-Defect product is of course impossible and Sustainable Tourism need not be that much of a Utopian idealism. While it is said that ICT can imbibe a qualitative improvement into the tourism system, it does not preclude vigil and proper planning at each stage of development. Tourism is more than ever an information sensitive industry which is greatly impacted by modern technological innovations such as the Internet, worldwide web, and electronic commerce. Unfortunately, these tend to widen the divide between developed and developing countries with the potential for siphoning much of the potential gains from tourism back to the developed countries. There is growing consolidation and centralization of the tools of the tourism trade among a few players. It is important to prevent the new medium turning into a tool for the powerful to reinforce the inequities of the past.

The informationalization of economy appears to be a significant aspect of tourism development as well as a valid instrument to reverse the inequalities of the social system. The most significant aspect of developmental debates associates to the abilities and skills of policy makers to access to reliable information and dataset. In this respect, with the rise of information society, long dormant asymmetries caused by the colonial period have become more evident. It is safe to say the effects of globalization and the social-economical changes generated by information technologies are pervasive. Dealing with more complex organizations based on information technologies is not only necessary but the only way to foster more resilient and sustainable communities.

The debate on 'technology for sustainable development' itself was launched by industry groups and business associations for whom it gave a concrete alternative answer to the 'anti-development' rhetoric on sustainability being propagated by non-governmental organizations and the 'new-left' civil society in general. They have produced numerous documents and policy statements on sustainable development outlining how the environment can be protected in the context of economic growth, freed-up markets, and industrial self-regulation. If somebody proclaims that aspects like the elimination of poverty and ecological sustainability are merged together in their design in synergic ways, it should not be accepted simplistically and acritically (Sheats, 2001). Critical investigations to the claims of interest groups have yet to be institutionalized, and it is the need of the hour to facilitate the growth of communities who compete for that pie since growth, both material and that of discourse, can be achieved only through such dialectical processes. The much-heralded movement attempting to invent and design different types of technologies that were more environmentally sound ('appropriate technology') if failed to influence the pattern of technology choice exercised by mainstream society and remained as a minority theme (Willoughby, 1990), one of the major reasons is that its development has mostly been via acritical monologs. Within the limitations, non-linear system dynamics of policy changes for sustainable pathways should be appropriated for the benefit of the disadvantaged. This, however, has been found to be extremely problematic on the minus side; it may be very difficult to determine reliably when and where to apply these policies, and how to evaluate their impact while the non-linear nature of social policies makes control by the power group (who are enjoined by elitist research communities) actually easier-it might take only a small push to engender a big change in the system. In the present case, the fear is that the movement for sustainable tourism through technological innovations might turn out to be a 'grand success,' but in inequitable ways 'brutally' suiting the dominant forces mostly located in the tourist generating regions of the West. Their power to manipulate public opinion through media tactics and woo international bodies of standardization and quality control to institute a costly and sophisticated technological regime is a cause of very serious concern for all those who want to see a sustainable tourism future for mankind.

KEYWORDS

- **computerized reservation systems**
- **COVID-19**
- **crisis**
- **destination marketing organizations**
- **development**
- **digital technologies**
- **information and communication technology**
- **technology**

REFERENCES

Avgerou, C., & Land, F., (1992). Examining the appropriateness of information technology. In Bhatnagar, S., & Odedra, M., (eds.), *Social Implications of Computers in Developing Countries*.

Boudreau, M. C., Loch, K. D., Robey, D., &. Straud, D., (1988). Going global: Using information technology to advance the competitiveness of the virtual transnational organization. *The Academy of Management Executive, 12*(4), 120-128.

Dimitrios, B., (2000). Marketing the competitive destination of the future. *Tourism Management, 21*(1), 97–116.

Dutta, S., Kwan, S., & Segev, A., (1988). Business transformation in electronic commerce: A study of sectoral and regional trends. *European Management 16*(5), 540-551.

George, B., (2002). *Techno-Politics in India: Technological Cannibalism and Survival Strategies*. Presented at the global business and technology association conference'02 Inabsentia held at Rome and later published in the conference proceedings.

Hall, P., (1996). *The World Cities*. London: Heinemann.

Ho, J., (1997). Evaluating the World Wide Web: A global study of commercial sites. *Journal of Computer-Mediated Communication, 3*(1). [On-line]. http://www.ascusc.org/jcmc/vol3/issue1/ho.html (accessed on 1 October 2021).

Hofacker, C. F., (2000). *Internet Marketing*. Dripping Springs, Texas: Digital Springs.

Hoffman, D. L., & Novak, T., (1996). Marketing in hypermedia computer-mediated environments: Conceptual foundations. *Journal of Marketing, 60*(3), 50–68.

Hoger, E. A., Cappel, J. J., & Myerscough, M., (1998). Navigating the Web with a typology of corporate uses. *Business Communication Quarterly, 61*(2), 39–47.

Kaye, B., & Medoff, N. J., (1999). *World Wide Web: A Mass Communication Perspective*. Mayfield Publishing Company, Mountain View.

Kunkel, J., (1970). *Society and Economic Growth: A Behavioral Perspective of Social Change*. Oxford University Press: New York.

Sheats, J. R., (2001). Information technology in sustainable development. In: Dorf, R., (ed.), *Technology, Humans and Society* (pp. 146–158). San Diego, CA: Academic Press.

Sterne, J., (1999). *World Wide Web: Integrating the Web Into Your Marketing Strategy*. John Wiley & Sons, New York, NY.

Tata McGraw-Hill Publishing, New Delhi, pp. 26–41.

Toffler, A., (1981). *The Third Wave*. Bantam, New York, NY

Werthner, & Klein, S., (1999). *Information Technology and Tourism: A Challenging Relationship*. Wien-Springer: New York, NY.

Willoughby, K., (1990). *Technology Choice: A Critique of the Appropriate Technology Movement*. Westview Press, Boulder: 12.

Yuan, Y., & Fesenmaier, D. R., (2000). Preparing for the new economy: The use of the internet and intranet in American convention and visitor bureaus. *Information Technology and Tourism, 3*(2), 71–85.

CHAPTER 13

The Lingering Quest for Authenticity in Tourism: Is Authenticity Really Dead?

ABSTRACT

Without any doubt, authenticity occupies a central position in the formation of tourist experiences and sensibilities. However, authenticity seems to be a concept very hard to grasp. Senior sociologists like MacCannell, Urry or Boorstin tried to define authenticity from different-if not contrasting-angles. This chapter discusses critically the philosophical essence of authenticity laying the foundations towards a new understanding of the phenomenon. This chapter is conceptually weaved to provide an over-view of the debates on authenticity in the new context of tourism whose boundaries are exceedingly being laid down by the all-pervasive forces of globalization and technological revolutions which were until now treated as variables exogenous to the tourism system. It takes stock of the historical development of the conceptualization of authenticity by research communities rooted in the various contributory disciplines to tourism and attempts to re-situate it in the light of the new realities. In this process, a few avenues for the conduct of future research in this domain are highlighted, too.

13.1 INTRODUCTION

'If a destination is not on the Web, then it may well be ignored by the millions of people who now have access to the Internet and who expect that every destination will have a comprehensive presence on the Web. The web is the new destination-marketing battleground and if you are not there fighting then you cannot expect to win the battle for tourist dollars'   *(Richer et al., 2000, p. 4).*

Touristic gaze is oftentimes conceived as the seeking after of authenticity (Cohen, 1972; MacCannell, 1973). Perception of the authenticity of the experience is an important mediating variable affecting tourist satisfaction. In fact, touristic space itself is structured to satisfy the desire for authentic experiences that motivate touristic consciousness (MacCannell, 1973). Slogans without the word 'real' or terms synonymous with it are atypical in mass tourism promotional devises. Most visuals employ metaphors traditionally or stereotypically associated with the authentic: the third-world countryside and its greenery, indigenous people in exotic costumes, and so on. And, no doubt, this is absolutely demand-driven.

One dominant characteristic of the present generation is the 'new elite' travelers who were born and brought up in the urban areas, among surroundings alienated and perverted, in the industrial landscape of the work-a-day world. While a cream of them happened to be fortunate enough to have listened from their grandparents or so stories of human life intermingling with fowl and brute in the idyllic, pristine countryside unaffected by the smoke and dust of heavy industries, for the vast majority, the only source influential in help shaping conceptions of authenticity is the all-pervading influence of the mass media complex. Given the quantum of impact the modern mass media has in shaping individuals' and society's conception of authenticity, an issue that is indeed worth exploring is the nature and characteristics of the media scripted authenticity: is there any ontological togetherness between the more traditional understanding of the term and its neo-modern variant, how subjective experiences are different when the gaze is for the media-constructed reality, what are the marketing opportunities and how marketers exploit them.

13.2   AUTHENTICITY: BUT WHERE TO LOCATE IT?

Philosophers from time immemorial have been puzzled by the riddle of the authentic: Is there something which is authentic and if at all the answer is yes is it possible to experience the same? Taking cues from far back in time, dominant schools of Indian thought declared that everything but the supreme spirit (*Brahman*) is an illusion (*Maya*); that it is ignorance (*Avidya*) which misguide us to believe that forms and relations of existence are real and that every soul strives for liberation (*Nirvana*) from this tangle of illusions. But such strivings fall short most often since human mind is too complexly illusioned to overcome.

Dominant epistemological designs of shifting paradigms from time to time have influenced the thinkers of society in their manner of approaching the issue of authenticity. If defining a concept means going back to the past and looking at the different stages and societies it has gone through, realizing the difficulty to explain the word authenticity is yet an element of its own definition. Trilling (1972) noted that even while it is so intrinsic to the tourism phenomena, authenticity is an ambiguous term that resists definition. In the study of tourism, authenticity can either be the authenticity of the observed tourist object or the authenticity of the tourist's first-person experience (Wang, 1999). The conception of authenticity has undergone three or more major shifts over the past 50 years, with objectivist framings giving way to social construction perspectives and, later, existentialist, and postmodern ones. The last one-third of the bygone century witnessed postmodernists, some of whom argued that nothing worth naming as authenticity is purely unqualified and that claims for authenticity should at best go into inverted commas (Urry, 1990). The increasing acceptance of the post-postmodern paradigm of critical realism as a bridge between the modernist and the postmodernist perspectives is another interesting development in the contemporary debates on authenticity. However, 'much of heat and very less light' remains the overall state of affairs. Notably, discourses, though they may have good internal consistency and rigor (postmodernists are humbler in not claiming even that much), lack proper transformational devises to make them communicable with those outside of them, including the marketing discourse which identifies, legitimizes, and capitalizes the purported 'objective' reality existing out there in the empirical world. While it is true that tourism practitioners conveniently employ the 'authentic' card to woo tourists to destinations, this apparent incompatibility between the generic social science and tourism marketing discourses poses serious difficulties in analyzing the issue in its comprehensiveness.

Marxian criticism posits that the more industrial the societies are the more alienated their population become. Since industrial societies in general have more surplus disposable income, most of the world's touristic flows naturally originate from the most alienated societies to societies that unfailingly preserved authentic forms of life. What is evident from this is the conceptualization of tourism as an exchange of the surplus disposable income with the authentic experience of the natives in the third world in an unusually asymmetric manner. Indeed, alienation is the most common denominator of mass international tourism as it creates its 'other,' the search for authenticity, which becomes a push factor and responds to the alluring call of tourism marketing enterprises offering authentic experiences in a world untouched by industrialization and its myriad evils.

Cohen (1988) says that the search for authenticity varies in direct proportion with the increasing level of alienation felt in a society. Tourists may go in search of unspoiled natives surrounded by landscapes of pristine beauty because these are absent in their advanced society (van Den Berghe and Keyes, 1984), but conditioned by those ways as determined by the forces of their social shaping. For the early McCannell (1976), authenticity in tourism products such as festivals, rituals, dress codes and so on can be determined straightly in terms of whether those are made or enacted by local people according to tradition. Such a position has every pitfall of inferring ontology from epistemological cues. While accepting the ultimate inability of an outsider to penetrate the destination culture, Boorstin (1961) points out that holidaymakers knowingly consume pseudo and contrived events to authentic cross-cultural encounters. Levy-Strauss (1989) writes that he is amazed at the will of tourists to believe the sacred fantasy as reality and to resist any other real as even potentially possible.

A comparative study to measure the perception of authenticity among visitors of 'The Rocks,' a historical neighborhood in Australia was made by Waitt (2000) which revealed important differences in the perceived level of authenticity related to gender, age, and place of residence. McKercher and du Cros (2002) argued that the Japanese happily accept faked tangible heritage assets. The tourist has become the symbol of a peculiar type of inauthenticity himself (Redfoot, 1984). Probably, as the old wisdom goes, truth grows inversely proportional to sacredness.

13.3   AUTHENTICITY AS AN INSTRUMENTAL VALUE

From the point of view of tourism marketing practitioners, authenticity is a unique capability that regions with spectacular attractions may be imputed as possessing that enhance value to their services thus help building an inimitable competitive advantage. Marketers believe that, in this way, they offer superior value to their customers. When it comes to marketing, authenticity, as Eco (1986) notes, is less historical and more visual. For the present-day man, there is no means whatsoever to peep into any higher reality than is provided by the media. For the present generation, the boundaries between fictional and real landscapes can be understood only in terms of the stories manufactured by the marketing communicators whose multi-media visuals echo as the tone of the implicit dominant ideology. Boorstin (1961), for instance, wrote about the inability of contemporary Americans to experience reality and they celebrate their idea of the real by participating in pseudo-events. Baudrillard (1988) notes that no object has an objective meaning than that implied by the messages communicated to get one familiarized with that object. It is not the material object that is consumed, but objectified signs: the idea of the relation. Thus, in order to become the object of consumption, an object has to become a sign. MacCannel (1989) seems to agree with this. He elaborates the idea of markers as the characterizing feature of any object as an object of consumption.

In the regular work life, increasingly, modern man lives in networked virtual (hyper-real) spaces where everything is a marker for something else. Tourism places are also being defined ever less in terms of geo-territorial integrity. Appadurai (1990) brought out 'deterritorialization' as one of the greatest existential concerns of our times. He noticed indigenous and tribal populations replicating Diasporas elsewhere in the world and enacting traditions in selective ways. Web-enabled deterritorialization is but a technology powered variant of this.

13.4   MEDIA 'MEDIATION' AS A KEY TO AUTHENTICITY

Thus, in the postindustrial era, marked by the forces of globalization and the Internet the idea of authenticity seems to have gotten with one dominant meaning provided by the mass media and any meaning incompatible with this is suppressed at its origin. This is not merely an analytical statement, and is true irrespective of one is positing the postindustrial reality as an epoch or as a different epistemic way of looking at events. Evident to anybody is the omnipresent layers of mass media representations engineered with a view to shape the human experience and to give individuals a sense of false confidence in these representations. These representations distort even those ideological and cultural distinctions that traditionally differentiated individuals in their touristic pursuit. In a post-structurally informed world, all ultimate meaning is illusory but the existential condition created by this necessitates people to individually and collectively create meaning. According to Cohen (1988), this is the same cause that gives birth to 'emergent authenticity' in tourism, essentially characterized by an ahistorical spatio-temporality.

Local communities living in destination areas may tend to use mass marketed images of them to weave an emergent worldview about their identity and position in the world and then enact it circumscribing their traditional roles, thus opening themselves up for the tourist gaze (Hobsbawn and Ranger, 1983). This, in addition, makes the demonstration effect, which is the tendency of locals to imitate tourist behavior (Smith, 1989), less artificial. Over and above, the capabilities of the new media facilitate the tourism firm and the individual traveler in co-manufacturing customized holiday meanings thereby enabling the firm to deliver 'mass-customized' services to each traveler by ingenious alternations in the marketing message suiting diverse customers and markets. Prototypically alienated tourists see in the media images of sites and attractions measures of their own experience (Sontag, 1977). By facilitating the construction of different readings, information, and communication technologies (ICTs) amplifies, facilitates, flexifies, and extends the tourist's opportunities, knowledge-base, and experience with no particular ontological taxation. For instance, research conducted by the author revealed that internet-based tour operators could successfully market the very same destination of Goa in India to diverse market segments as diametrically different interpretative readings after grabbing and assessing the customer profile from the Net and other sources (George, 2003). It is an interactive game of co-manufacturing reality beneficial to both the customer and the marketer. Brown (1993) attempted to capture this tripartite nature of the production process of symbolic meaning involving the buyer, seller, and the media by introducing the idea of a 'pluri-signified product.'

In a situation like this, the business success of an individual who host customers at a destination will depend up on her mettle to enact and shift among as many the aforesaid multiple readings of 'reality' as per the demands made by the divergent segments of customers (Ashworth and Tunbridge, 2000). This may make customer relationship management in the future painfully dramaturgical. Or, technology should effectively mediate the guest-host interactions so as to create an impression of the opinionated reality being enfolded uninterruptedly. The proliferating virtual sites on the web far outnumber any increase in the physical sites existing to match their claims, which complicates the matter further. The upper cap of the potential tourist sites on the web is infinity, but there is no mandatory requirement that for this to happen the geographical territories need to be equal in number.

13.5   FLEXI-PRODUCTION OF MEANINGS: AN OPPORTUNITY IN DISGUISE

The possibility of simultaneous production of multiple meanings for catering to the needs of diverse markets is laced with a golden opportunity too. In the marketing terminology, alternation in the fourth P of marketing, i.e., promotion is increasingly being thought of as the first and implemented as the most frequent option in any marketing strategy by destinations. The capability of the new media, if exploited ingeniously, may lead unto more sustainable development of the destinations since alternations in the first P, i.e., product, is now considered rarely as an option to meddle with according to the whims and fancies of the short-term market demand. But, the media-controlled (re)construction of meanings is also a complex political process in which many voices are suppressed while a few amplified and occasionally some new voices integrated from nowhere. The place images that emerge to represent a place as a tourist destination, for instance, incorporate only very few (ab)original elements of that place from amongst the multitudes of understandings held traditionally. These multiple meanings of place are continuously negotiated and contested by residents and visitors to the destination. Going back to our previous studies (George, 2003), certain versions of Goa's place identity were heard more loudly as apprehensions in the local public forums while those very same versions arouse spirited interest in the international mass tourist market for the destination, thus both discourses thrive one upon another.

13.6   UNDERSTANDING TOURISM: THE MEDIA-WAY

If the new media is negotiating with the tourist for the marketer in the process of constructing and consuming touristic experiences, is there a better opportunity to theorize the touristic experience in the context of globalization and ICT revolution than critically looking at the very same media, asks MacCannell (1999). The position taken by him is that the marketer's message is not something that can be disaggregated from the media that is used to transmit it. Note that the analysis of touristic authenticity is made problematic since in the media generated world everything is a heuristic to interpret something else and no image can be finally determined to be closer to the original than the others. This is because even the so-expressed original is an image of something else. Hence, analyzes get into a mode similar to that of literary criticisms that have validity only within the bizarre linguistic reality structured by the 'montage' (Baudrillard, 1983) and hence become self-referential. McLuhan (1964), the fountainhead of modern mass media studies, while trying to make a modernist rational picture of the media and the message did not appreciate this aspect well anywhere in his studies. (Of course, McLuhan's conception of media is much wider in scope than we envisage in the present analysis).

'The role of advertising therefore consists of transmitting intact to the periphery… a model in the form of messages to which the mass media contribute the necessary force, while the larger social milieu verifies that the messages and the goods follow the expected norm' notes, Thurot and Thurot (1983). Again, the issue at stake is that the 'larger social milieu' has virtually no way to 'verify' unmediated by the media and the 'expected norm' itself is not independently developed without the influence of the grand media complex.

The supremacy of media to deliver what the receiver expects and in the continuing process of feedback loops orient his expectations to what it can provide is increasingly evident in the applications of modern ICT and has already been discussed in this chapter. The point is that, as Said (1978) noted, text acquires greater authority and use than the actuality that it purports to describe. In the tourism context, this means that, when tourists report of their recent tour as just a mockery and cheating it just connotes that their experience was something different from what the media had offered it would be. 'What you expect is what you get,' problematizing even the basic push factors for touristic search such as 'alienation.' For the youth of our time, alienation itself is what is set as alienation by the media. Over and above, solutions for triumphing over alienation also has to necessarily come from the media. In other words, the same media that generate the 'other' will automatically, as a logical necessity, generate clues to its opposite, the non-touristic identity. Yet, as an economic rule, commercial media do not for itself help mold one's self, since demand for marketed commodities can be generated only by pointing out what is lacking in one rather than what one is having.

13.7 THE PROGRESS HYPOTHESIS: A REAPPRAISAL

What does the above discussion say of the 'progress hypothesis' involved in the historical development of mass media from print through radio to the Internet? It could be argued that paralleling with technological development the consumer was being made more and more passive and disconnected. While keeping him in the illusion that he is permitted more and more access, uninterrupted connectivity, and say in determining the content and form of the informational feast dispatched to him and luring him to indulge in 'depth,' media claims about the provision of authentic information became ever less verifiable. Radio as a medium, for example, transmitted stories to a mass audience, but it resembled more like the grandma's tales demanding the listener to imagine and think. And it did not preclude chances for the audience to verify the truth of what has been told via other sources, mostly direct verification, and inter-personal means. And in fact, there existed enough 'free space' not capitalized by the mighty mass media through which one could conduct corroboratory search. Television gave colorful images and scope for much deeper involvement but taxed the important elements of attention and thought from the viewer that would otherwise have been spared for reflection and verification. Along with this the uncapitalized free space for verification also became narrower than ever before. Latest, the Internet's interactive options and 'coproducibility' ensured that any gap in a narrative or interpretations whatever required to appreciate the story could as well be extracted from the help menus, FAQs, related links into other web sites and the likes. The effortlessness for this was too tempting for the user. Consequently, the option was to verify all claims within the labyrinths of the cyberspace.

Paradigmatic shift that began with the new media is that while the traditional media extended the presence of the news source to the doorstep of the audience, the Internet did the opposite by facilitating an extension of the presence of the 'browser' to the homes of corporations. First, you reach the homepages of organizations from where, links take you to inside rooms and, in this process, complex algorithms map your 'cyberology' to see if they have products suiting you or if you have anything to suit their offerings.

Habermas's (1989) critique of the disintegration of the 'public sphere' in the capitalistic society can posit this set of descriptions in to a wider sociological theory. According to Habermas, the public sphere has been transformed from its original content-focused openness, into something very hollow and superficial. The world in which we live day-to-day has been colonized by the market economy and legal bureaucratic regulation. Habermas is distressed with the power and influence held by the mass media. 'Whereas the press could previously merely mediate the reasoning process of the private people who had come together in public, this reasoning is now, conversely, only formed by the mass media,' he reflects. In other words, public opinion in today's world is more a resultant of corporate and government manipulations, both alienated from citizens, than of the interactions among private citizens. Society is not controlled by critical reason and the political choice of the public. Publicity has become a commodity that governments and corporations bring into play to protect their interests and spectacular image. The public has been depoliticized, and the public sphere deformed.

The original ideas of Habermas applied into the realm of the new media while attempting to answer whether there are new kinds of relations occurring within it [the Internet] suggest new forms of power configurations in the society and reactionary consciousness among private citizens. The Internet, thus, becomes a systematic denier of the public sphere. While taking a position like this, the author does not contest those subjective answers to whether this is debility or empowerment will still depend up on one's ideological orientation.

13.8   VARIATIONS WITHIN THE 'MASS'

The above exposition should make it clear that tourists inhabit hegemonically scripted and mass-media-mediated discourses. But the influence is not cloned for everyone in any singular way. For instance, Cohen (1979) himself admits that there are tourists (not 'The Tourist') who seek neither fantasy nor escape from reality. There are also those who actively challenge the staged reality by getting into the roots (Redfoot, 1984). Arbitrarily based on the concepts of 'staged authenticity' and 'tourist space,' Cohen (1988) describes four types of touristic situations. The core in this conceptual framework is authenticity, e.g., what is the tourist's impression of a scene. He argues that the breath of authentic traits necessary to satisfy the tourist will, in turn, depend on the depth of the touristic experience to which each individual tourist aspires. (But it could still be argued that each category of tourists participates in its own forms of inauthenticity (Naoi, 2004).

Note that it is not the individual's volition to become a particular tourist type oneself (Nasar, 1998). Individual freedom is constrained and perpetually modified by powerful discourses extrinsic to him. A set of core values is known to ferment in the historical development of any interdependent community. This evolutionary wisdom acquires more and more independence or portability in course of time and begins to actively shape the perception of individuals living in that community (Inglehart and Baker, 2000). Different societies condition the individual psyche differently in the matter of how and what to look for in cultural artifacts. Even though it is comfortable to learn that cultural variables as contrasts have ceased to be true in the light of rapid waves of globalization, serious studies from elsewhere points out that cultural differences and their 'prismic' effect upon attitudes and behavior have only become significantly more visible than ever before due to these developments (Huntington, 1996; Landes, 1998).

13.9 CONCLUSION

This last chapter provides a fresh insight on the ethical dilemmas of authenticity while lading the foundations towards a new understanding of the globalization process. To date, some global forces, which are exogenous to the tourist system, threatens the relation between hosts and guest. For that, scholars are urged to rethink new boundaries for the concept of authenticity that have been widely treated in different disciplines. In this vein, the chapter gives attention to the importance of future research in the constellations of authenticity-related issues.

As noted, the issue of authenticity is fundamental to any learned sociological understanding of the tourism phenomenon. Authenticity as a topic of study definitely deserves serious attention in a world that is struggling with the challenges of novelty, diversity, and chaotic cultural shifts. We must candidly accept that, in spite of the unique nature and characteristics of authenticity as applied to tourism, only a few special semantics that are sympathetic to its unique nature have been evolved reasonably well. Directing a critical mass of research activities towards this end is sine qua non for tourism to have a separate space among wider and disparate discourses on authenticity as well. More worrying is the fact that research commissioned by commercially interested sections advances marketing knowledge on how to capitalize by commoditizing authenticity at an ever-greater pace (Ashworth, 1991; Sack, 1992) and such private knowledge is couched in ideological and propagandist terms prior to serving for public consumption (Hewison, 1987). To resist this sort of an uncritical and unbalanced growth is contingent upon each one of us who visualizes a just, free, and informed world.

KEYWORDS

- **consumption**
- **cyberology**
- **globalization**
- **hermeneutics**
- **philosophy**
- **the death of authenticity**
- **tourism**

REFERENCES

Appadurai, A., (1990). Disjuncture and difference in the global cultural economy. *Theory, Culture and Society, 7,* 295–310.

Ashworth, G. J., & Tunbridge, J. E., (2000). *The Tourist-Historic City: Retrospect and Prospect of Managing the Heritage City (Advances in Tourism Research Series).* Pergamon, Oxford.

Ashworth, G. J., (1991). *Heritage Planning: Conservation as the Management of Urban Change.* Geo Press: Groningen.

Baudrillard, J., (1983). *Simulations* (trans. by P. Foss, 1983) Semiotext (e), New York, NY.

Baudrillard, J., (1988). The system of objects and consumer society. In: Mark Poster (ed.) *Selected Writings.* Verso London.

Boorstin, D. J., (1961). *The Image: A Guide to Pseudo-Events in America*. Athenaeum, New York NY.

Brown, S., (1993). Postmodernism: The end of marketing. In: Brownlie, et al., (eds.), *Rethinking Marketing: New Perspectives on the Discipline and Profession* (pp. 1–11). University of Warwick: Warwick.

Cohen, E., (1972). Toward a sociology of international tourism. *Social Research, 39*(1), 164–182.

Cohen, E., (1988). A phenomenology of tourist experiences. *Sociology, 13,* 179–201.

Cohen, E., (1988). Authenticity and commoditization in tourism. *Annals of Tourism Research, 15*(3), 371–386.

Cohen, E., (1988). Tradition in qualitative sociology of tourism. *Annals of Tourism Research, 15*(1), 29–46.

Eco, U., (1986). *Travels in Hyper-Reality*. Harcourt, Brace, Jovanovich, San Diego.

George, B. P., (2003). *Exogenous Innovations and the Reinvention of Travel Intermediaries: Theoretical Considerations and Empirical Findings* (pp. 3–12). Pre-PhD thesis, Goa University, India (also published in DC journal of management.

Habermas, J., (1989). *The Structural Transformation of the Public Sphere: An Inquiry into a Category of Bourgeois Society* Polity Press, London.

Hewison, R., (1987). *The Heritage Industry: Britain in a Climate of Decline*. Methuen, London.

Hobsbawn, E., & Ranger, T., (1983). *The Invention of Tradition*. Cambridge University Press, Cambridge.

Huntington, S. P., (1996). *The Clash of Civilizations and the Remaking of World Order*. Simon and Schuster, New York, NY.

Inglehart, R., & Baker, W., (2000). Modernization, cultural change and the persistence of traditional values *American Sociological Review, 65,* 19–51.

Landes, (1998). *The Wealth and Poverty of Nations: Why Some are so Rich and Some so Poor*. W.W. Norton & Company, New York, NY.

Levi-Strauss, C., (1989). *Tristes Tropiques*. Picador, London.

MacCannell, D., (1973). Staged authenticity: Arrangements of social space in tourist settings. *American Journal of Sociology, 79*(3), 589–603.

MacCannell, D., (1999). *The Tourist: A New Theory of the Leisure Class*. Schocken Books Inc, New York, NY.

McKercher, B., & Du Cros, H., (2002). *Cultural Tourism*. The Haworth hospitality press, New York, NY.

McLuhan, M., (1964). *Understanding Media: The Extension of Man*. McGraw Hill, New York, NY.

Naoi, T., (2004). Visitors' evaluation of a historical District: The roles of authenticity and manipulation. *Tourism and Hospitality Research, 5*(1), 45–63.

Nasar, J. L., (1998). *The Evaluative Image of the City*. Sage, London.

Redfoot, D. L., (1984). Touristic authenticity, touristic angst and modern reality. *Qualitative Sociology, 7,* 291–309.

Richer, P., et al., (2000). *Marketing Destinations Online: Strategies for the Information Age*. World Tourism Organization Publications, New York, NY.

Sack, R. D., (1992). *Place, Modernity and the Consumer's World*. Johns Hopkins University Press, Baltimore, MA.

Smith, V., (1989). *Hosts and Guests: The Anthropology of Tourism* (2nd edn.). Philadelphia University Press, Philadelphia, PH.

Sontag, S., (1977). *On Photography.* Farrar, Strauss, and Giroux: New York.

Thurot, J., & Thurot, G., (1983). The ideology of class and tourism: Confronting the discourse of advertising. *Annals of Tourism Research, 11*(3), 173–190.

Trilling, L., (1972). *Sincerity and Authenticity.* Harcourt Brace Jovanovich Publishers, New York, NY.

Urry, J., (1990). *The Tourist Gaze.* Sage, London.

Van, D. B. P., & Keyes, C., (1984). Tourism and recreated ethnicity. *Annals of Tourism Research, 11*(3), 343–352.

Waitt, G., (2000). Consuming heritage: Perceived historical authenticity. *Annals of Tourism Research, 27*(4), 835–862.

Wang, N., (1999). Rethinking authenticity in tourism experience. *Annals of Tourism Research 26*(2), 349–370.

CHAPTER 14

Complexity, Uncertainness, and Tourism: Tourist Consciousness

ABSTRACT

Tourism and hospitality are sensitive to bad advertising and risky climates of businesses. In a hyper-globalized world, which is in change all the time, tourism has serious problems to forge the necessary political stability to prosper. The argument in this chapter goes on to say that governance has triggered a hot debate respecting to the opportunities of developing economies to adopt tourism as a successful form of poverty alleviation. To a closer look, the development theory-as well as the notion of governance-is certainly based on long-dormant geopolitical discourse forged in the days of European colonialism. The tourist consciousness, a neologism used in this chapter, speaks us of the set of narratives, verbal construes, stereotypes, and discourses oriented to symbolize the "Non-Western Other." The chapter ends with a useful encounter between two senior sociologists who have worked on the impact of visual modernity and poverty, R. Tzanelli and B. Freire Medeiros. Both agrees with the thesis that tourism embellishes poverty through the articulation of a spectacle, and in so doing, the industry perpetuates the conditions of exploitation in the global slums.

14.1 INTRODUCTION

It is tempting to say that tourism not only is sensitive to bad advertising but also is directly associated with risk avoidance. The term governance has been enthusiastically adopted to promote more resilient destinations. The political stability as well as the cultivation of democratic institutions seems to be vital for the survival and expansion of the tourism industry. As discussed in the different earlier chapters, which take part in this book, countless risks ranging from terrorism to political inter-ethnic conflicts-without mentioning ethnic cleansing or wars-have placed the industry in jeopardy. The present chapter interrogates furtherly on the problem of governance and politics to form what specialists dubbed as "tourist consciousness" (Li, 2000; Bates, 2003). Over recent years, tourism, and hospitality have notably grown. At the same time, tourism has successfully revitalized global commerce and different developing economies world-wide (de Kadt, 1984; Sugiyarto, Blake, and Sinclair, 2003). Having said this, the job multiplication factor associated with the economic growth occupies a central position in the political rhetoric. To support their agenda and administration, professional politicians allude to the economic benefits of the tourism industry. A sign of good governance is certainly based not only on economic success but in domestic consumption, which is a clear indicator of governance. The act of traveling is more than complex than a mere physical displacement; so, to speak it exhibits a complex ideological imprint which mediates between lay-citizens and their institutions. By traveling people legitimate their political authorities while internalizing "the importance of being part of a privileged society" (Gossling and Hultman, 2006; Hultman and Hall, 2012). Presidents allude to tourism to validate their administrations (Scheyvens, 1999; Sofield, 2003). Not surprisingly, tourists often valorize their societies considering they live in a democratic, liberal, and prosperous atmosphere where human rights are protected. The Western social imaginary punctuates that tourism does not prosper in undemocratic societies or communities devastated by inter-ethnic conflict. To some extent, tourism, and economic success seems to be inextricably intertwined. With the rise and expansion of globalization, the local low scale economies have set the pace to a complex and uncertain world, where global risks place the industry between the wall and the deep blue sea (Tarlow, 2006, 2014). What is more important, the recent virus outbreak originated in Wu-Han, China (known as COVID-19) has para-lyzed the global commerce changing radically not only tourism but also the geopolitical relations as never before. As Geoffrey Skoll (2016) puts it, the capitalist system has expanded and in so doing, the fear which is its ideological core has been flourished to the four corners of the world. The history of the US is plagued with political discourses finely orchestrated to demonize the "Non-Western Other." Centered on a culture of fear where this "Other" is neglected, capitalism creates, packages, and disseminates a "Spectacle on Terror" for the working class to be passively dominated.

As the previous argument is given, policy-makers in tourism devoted considerable efforts and time planning more resilient destinations. Recently, some voices have alerted on the problems of governments to protect foreigner tourists or anti-tourist manifestations. Besides, some violent episodes in Europe and the US have captivated the attention of specialists (Milano, Novelli, and Cheer, 2019). In his book, *Tourism, Terrorism, and the end of Hospitality in the West,* Korstanje (2018) argues that terrorism is mining the social trust of capitalist societies destroying one of its sacred laws, hospitality. The Wall proclaimed by Donald Trump in the US; the incipient Islamophobia in Europe adjoined to the multiplication of racial riots are good examples that prove hospitality-at least as we know it-is dying. Part of this problem has eloquently addressed in different chapters of the present book. In an ever-changing and uncertain world, the future of tourism remains an open question. The question whether Emmanuel de Kadt imagined top-down planning as an instrument for tourism development has been changed by a new paradigm where uncertainness- and of course bottom-up planning-plays a leading role to design more sustainable destinations.

14.2   INITIAL DISCUSSION

Without any doubt, the tourism industry can be valorized as a vehicle towards peace (Litvin, 1998; Farmaki, 2017). Under some conditions, tourism creates more tolerant and open societies but sometimes it is the exception to the rule. As Comaroff and Comaroff (2009) brilliantly described in their book *Ethnicity Inc.* Tourism allows a rapid reconstruction in communities which have been historically relegated from the wealth distribution but paradoxically, it revives long-dormant conflicts, which unless dully regulated, may lead towards the escalation of violence even to ethnic cleansing. Some ethnic minorities situated the peripheral positions have much to gain with tourism, but sooner than later they are forced to pay higher taxes, awakening violent separatist reactions. At the same time, tourism leads some minorities to engage in their national heritage, no less true is that paradoxically hostility against neighbors or foreigners aggravates. Hence the pathways to sustainable governance are far from being clear. Although the complexity theory has recently arrived in tourism research, it came to stay. At a closer look, complexity theory has been adopted to confront and override some already-existent paradigms. Above all, it defies the ideals of sustainability and rational planning that marked the pace of tourism research for years. For those scholars who toy the belief, rational planning equates in a more robust and sustainable destination is far from being validated in this global and uncertain world. In this vein, governance-far from a panacea-sets the agenda of governments in a world where uncertainness and risk prevail. Of course, the specialized literature suggests that tourism has positive and negative effects on the community (Jafari, 1994). Originally, the all-inclusive tours offered a protective cocoon for the foreigner tourist. The Bubble model, where host and guests kept low interaction, was at the forefront of the outset of tourism research (Burns, 2004). After Luxor Massacre, where almost 58 tourists were killed in cold blood by a radicalized terrorist cell in Egypt, experts realized the sustainability of tourism cannot be taken for granted unless host guests interaction is encouraged. After all, the evidence showed that the Bubble model seemed not to be sustainable in the time without local hospitableness (Mansfeld and Pizam, 2006). In this respect, Sevil Somnez acknowledges that the scourge of terrorism has cemented the evolution and maturation of tourism research, allowing further coordination between management and Academia. Doubtless, terrorism, and political instability are two sides of the same coin. Those nations marked by violence and political instability have further probabilities to suffer terrorism than democratic societies, Somnez concluded (Somnez, 1998). Of course, she was unfamiliar with the 9/11 and the successive attacks in Europe from 2001 to date. To wit, terrorists look to harm foreigner tourists not only to create chaos but to humiliate the hosting nation. Serious diplomatic issues emerge just after a terrorist attack, above all when the hosting state failed to protect foreigner tourists. After the attacks in Bali (2003), Australia claimed internationally for Sri Lanka to the next steps to locate and trial the responsible terrorists. Not only tourist destinations but also geopolitical relations are sensitive to terrorism and political violence (Pizam and Fleischer, 2002; Liu and Pratt, 2017). The example of international terrorism offers a fertile ground to the creation and imposition of policies and protocols aimed at achieving a climate of stability to boost local economies. Governance occupies a central position in the social imaginary as a valid instrument towards a more resilient industry.

14.3 GOVERNANCE AND TOURISM

The term governance was originally coined in the 90s decade to denote the efficiency in the state policy-making to intervene in those risks which put its legitimacy in danger. One might speculate in ever-changing geography the public and private spheres should work together to enhance the "local governance." As a dogmatic doctrine, governance takes different meanings depending on the ideological academic paradigm of the epoch. In tourism and hospitality, the governance theory couples with sustainability, development, and poverty relief. As Bramwell and Lane (2011) suggest, governance is the touchstone of rational planning simply because it corrects the material asymmetries generated by the capitalist system and tourism consumption. While sustainability appears to operate to long-run, governance works effectively in a short-run. Both concepts, experts say, are inevitably entwined.

In consonance with this, Giana Moscardo (2011) alerts that the theory of governance should not be limited to an economic dynamic nor a centralized decision-making process in the executive branch, rather governance is reached by different stakeholders and agents. The obtained knowledge, not politics, is of vital importance to reach a state of durable governance in tourism. Arturo Escobar (2011) gives a fresh insight in this direction. Governance results after genuine dialog and negotiations where different voices converge. After the WWII ended, the discussion was given to the policies oriented to poverty relief. In so doing, tourism showed to multiplicate jobs having positive durable effects in the economy. However, once the industry expanded, the world became less predictable and stable. The governance theory replaced the established development theory. Today's policymakers reach consensus in holding that asymmetrical societies have fewer opportunities to offer efficient solutions to the problems of governability. The doctrine of governance rests on a philosophical dilemma, as Hall (2011) eloquently reminds. The quest for profits has been illuminated the broad strokes of the specialized literature for decades, but far from being a solution, it represents the problem. While preliminary the tourism industry expands, local economies are buttressed. Though the productivity of tourism industry revitalizes not only economies but energize state administration, no less true is that in a second facet, the natural environment is negatively affected creating the conditions to an ecological crisis. Here, the literature divides into two clear-cut poles. On one hand, those studies which focus on governance as the consequences of policies and the interaction of different stakeholders to set collective goals. On another hand, other works remark in profits and wealth-maximization as two key factors to strengthen governance (Beritelli, Bieger, and Laesser, 2007; Volgger and Pechlaner, 2014; Blasco, Guia, and Prats, 2014; Baggio, Scott, and Cooper, 2011). What is more important to discuss here, unlike in other disciplines, in tourism the term governance is commonly applied to poverty reduction, economic success and development theory. But precisely it says little about what is wrong with the development theory? a point which remains unquestioned in the constellations of tourism fields.

14.4   WHAT IS WRONG WITH DEVELOPMENT THEORY?

Although this is a question which was not addressed in tourism, it was already formulated and widely studied by cultural anthropology. Much has been written on the development theory and its effects on economies but unfortunately, less has been said about the ideological nature of the term which is originally coined by US former president Harry Truman in 1947. In that speech, Truman called for rich (developed) nations to bail out under developing nations. Immediately to this speech, the word was cut in two: developed and underdeveloped countries. For that, it is safe to say development theory interlinks directly to the narratives circulating in the Cold War.

The end of the Second World War left Europe into the brink of bankruptcy with serious human and financial losses. In Truman's mind, the US should take the lead marking the borders between a civilized and democratic world, formed by richer and liberal nations which should morally assist to poorest economies of the planet. As he thought, with this financial aid adapted to an educational program, those underdeveloped economies would have the opportunity to achieve a more maturate stage of production reaching many of the benefits of consolidated and leading economies. The triumph of the US was not only marked by the so-called cultural superiority of American lifestyle-if not its mainstream cultural values-but with the moral urgency to export this successful model to the world (Gardner, 2002). From that moment on, Esteva comments, central countries-most of them situated in the Global North-articulate countless financial programs to promote the development and political equality in the Global South, but without practical results. Not only decades of development theory have not generated richer countries, but paradoxically, solicitant countries impoverished notably following the programs of the World Bank or IMF. The development theory condenses the cultural values of the US in a type of Western nostalgia for protecting the more vulnerable citizens, an idea inscribed in the European colonialism. Esteva exerts a radical critique against the development theory which leads us to think it is ethnocentric at least in its application. He holds polemically that development theory has ushered the Global South in a nightmare reinforcing the old dependency between having and have-nots (Esteva and Prakash, 1998). Despite the volume of money bailed out in the system and the international loans issued without any control by the IMF, developing countries are poorer than decades ago.

Proponents and exegetes of development theory overtly claim that no all countries can reach a developed state of production simply because of cultural incompatibilities such as a blood legacy of inter-tribal wars, successive coups or political instability or simply political corruption (Escobar, 2011). In a landmarked book which entitles *Development and Social Change,* Phillip McMichael traces back the evolution and limitations of the theory. And of course, he did with extreme acutance and clarity. McMichael is moved to describe the ebbs and flows of development theory in the West since the 50s decade. He masterfully dissects the global (ideological) framework that allows the implementation of financial aids in the cold war period. The US urged to re-colonize a new world once Nazi Germany was defeated. The red scare associated with the Cold War paved the ways for the implementation of financial support to countries devastated by the war. Here an additional problem arises, the colonial empires castigated by the material losses and an economic downturn faced serious problems to keep their colonies under control. A set of different political claims oscillating from economic benefits to democracy and the self-representation prompted the over-seas territories to proclaim their independence from the European metropolises, in the grim days of what historians known as "decolonization process." This happened because the European Empires never shared with the colonies the financial benefits of democracy and the ideals of check and balance powers. Colonizers and colonized were twined by the needs of adopting democracy as the main form of political organization. In a context of change and disintegration, the development theory continues with the ideological narrative of colonialism, keeping the liaison between Europe and its former colonies. The introduction of colonialism sets a state of submission ideologically organized to exploit the "non-Western Other." Sooner than later, aborigines realized the double moral standards of Europeans who boasted of their democratic spirit while oppressing their colonies. This ambivalence was filled by the development theory once the independence was more than a simple dream. The development theory served not only to placate the Soviet influence but as a new cultural project for oiling the mechanisms of control over the periphery. McMichael calls attention to the impact of development theory on food dependency. The third world limited to export raw-materials and commodities to the developed economies while elaborated products were returned. The globalizing process gradually occupies the same role of the development theory when serious questions about the legitimacy of IMF and World Bank surfaced. The lack of legitimacy of these organizations was given by the successive failures in administering development-related programs in the Third World. Even in Africa, the effects of the development theory resulted in inter-tribal wars and ethnic cleansing. As the previous argument is given, McMichael argues convincingly that globalization was a success in expanding thanks to the lack of protective barriers of the Third world where the capital investors were welcomed. This, in consequence, provoked two alarming situations. An increase in unemployment and the decline of unionization in the North was accompanied by the arrival of international business corporations seduced by the low-cost of workers in the South. The doctrine of "free enterprise" was presented as a superior ladder in the evolutionary process. Each state should adopt a specialized role in a much wider "world factory" where some provide with the raw-materials and others with elaborated products. This trends which characterize the 90s decade created a new asymmetry between skilled (located in the first world) and under-skilled human resources (situated in the periphery). The recession produced by oil-embargo pressed First World to borrow a massive influx of money to the Third world, but now it will be carefully selected by two organisms, GATT, and WTO. Both curtailed the protective measure of local economies by consolidating of a new model which combined the reduced public capacity with the needs of governance. If Nationalism showed the importance of nation-state to protect the citizen from Market's arbitraries, now neo-liberalism focused on the inefficiency public administration to regulate the economy.

"In short, the making of a free trade regime reconstructed food security as a market relation, privileging, and protecting corporate agriculture and placing small farmers at a comparative disadvantage. Food security would now be governed through the market by corporate, rather than social criteria" (p. 136).

In a nutshell, globalization stimulated the profit accumulation centralizing the control beyond the authority of nation-states. Private corporations amass a whole portion of food production and distribution in the world while the conditions of poverty and slumming are notably increased in the food-producing countries. The auspices of governance seem not to be pretty different than the original sin of development theory. It portrays an idealized future somehow designed by the ideals of European culture which is superior to other cultural forms, based on the ideas of free-trade, democracy, and liberality. In this exchange, poor nations have the opportunity to overcome the next stage of evolution, but nobody tells at what cost. Free trade and democracy are presented as universal (desirable) values all cultures and nations should adopt. Of course, the failures and limitations to achieve more developed economies are never at the hand of development theory, rather it is a clear consequence of the "Non-Western Other's" cultural background. The figure of culture, here, plays a crucial role in marking the borders between civilized and uncivilized nations. To what extent, tourism helps in creating more resilient destinations that mean in stable conditions of production for governance but this does not happen in all cases. Under some conditions, development, and governance reanimate long-dormant discourses or conflicts in the developing nations.

14.5 CONCLUSION

Since its adoption, the term governance has influenced many scholars and researchers. With a focus on poverty relief and economic production, governance gave the hope for policymakers to struggle against poverty of certain ethnic minorities. The turn of the century witnessed the rise of old conflict and the radicalization of some terrorist cells. This was a fertile ground for the adoption of governance theory to explain the benefits of tourism in communities. As an instrument for durable peace, tourism consolidates democratic institutions and prosperous economies. As debated in this chapter, exegetes of governance-like development theory in the former century-are incognizant with the ideological nature of the doctrine as well as the center-periphery dependency. Postmodern sociology provides a timid acknowledgement on the limitations of governance theory. What is more important, there are real contradictions in the ways the poverty alleviation is conceived. We shall end this chapter synthesizing the main points of discussion between two senior sociologists interested by the role of tourism in the poverty relief, Rodanthi Tzanelli and Bianca Freire Medeiros. The convergence of both argumentations throws light on a clear answer revolving around the governance theory.

Tzanelli proffers a more than an interesting model to expand the current understanding of governance and tourism consumption. Per her viewpoint, the tourism industry produces ideological allegories which are negotiated and accepted by the agencies. These allegories-based on pseudo-historical facts-are imagined to meet the needs of a global consumer: the tourist. Some peripheral cultures internalize their so-called inferiority respecting to the European matrix through cinema, and tourist consumption. There is a great distortion between the gazed and the gazer. Today tourists are natives, and natives turn in tourists in a digital screen. As Tzanelli notes, the main problem of poverty is not the economic submission of some classes, but the creativity of capitalism which evolves through *creative destruction*. Artists and creativity worked hard to draw a utopian landscape shared and gazed by a broad audience worldwide. In this world, the disaster looms while interrogating us. The creative destruction allows the revitalization of social and economic structures, that represent an obstacle to the advance of capitalism. Those first world tourists who move to gaze slums not only are unfamiliar with the responsibilities of their states in the colonial period but also perpetuate the legitimacy with their state by a morbid fascination with the Other' pain (Tzanelli, 2013a, b, 2016). At the time, slumming, and slum tourism are commodities to exchange in a liberal marketplace, poverty tends to be perpetuated. This is the point of start for Bianca Freire Medeiros, who in his book *Touring Poverty*, revises the classic Marxian axiom that locals are ideologically dominated and exploited by tourism. In some conditions, locals take the advantage of the positive impact of tourism to control and exclude others. The access to capital is of paramount importance to organize slum-tours in the favelas (Brazilian slums). The marginality of favelados, as well as the stories of narco-guerrilla and police abuse, is offered as a cultural spectacle to foreigner tourists who visit slums in Rio de Janeiro Brazil.

Through tourism, locals echo their demands which often are not met by the government. This auto-engagement with community empowers favelados while perpetuating the poverty conditions. Favelados introduce tourism to improve the community's governance but in so doing, they involuntarily commoditize poverty as a good to be exchanged. In a world where poverty is a spectacle the probabilities to reduce it are simply low. Last but not least, this chapter discussed the problem of governance and its effects on local communities. For Tzanelli, the term denotes repressed discourses intended to legitimate a bloody colonial past but for Medeiros, the material inequalities of society are given by the logic of spectacle, tourism romantically evokes: this is the tourist consciousness.

KEYWORDS

- **development theory**
- **globalization**
- **governance**
- **mobilities**
- **post modernity**
- **poverty**
- **tourism industry**

REFERENCES

Bates, C., (2003). Hotel histories: Modern tourists, modern nomads and the culture of hotel-consciousness. *Literature & History, 12*(2), 62–75.

Beritelli, P., Bieger, T., & Laesser, C., (2007). Destination governance: Using corporate governance theories as a foundation for effective destination management. *Journal of Travel Research, 46*(1), 96–107.

Blasco, D., Guia, J., & Prats, L., (2014). Emergence of governance in cross-border destinations. *Annals of Tourism Research, 49*, 159–173.

Bramwell, B., & Lane, B., (2011). Critical research on the governance of tourism and sustainability. *Journal of Sustainable Tourism, 19*(4, 5), 411–421.

Burns, P. M., (2004). Tourism planning: A third way? *Annals of Tourism Research, 31*(1), 24–43.

Comaroff, J. L., & Comaroff, J., (2009). *Ethnicity, Inc.* University of Chicago Press, Chicago, IL.

De Kadt, E. J., (1984). *Tourism: Passport to Development? Perspectives on the Social and Cultural Effects of Tourism in Developing Countries (No. 338.4 KAD).* New York, World Bank-UNESCO.

Escobar, A., (2011). *Encountering Development: The Making and Unmaking of the Third World* (Vol. 1). Princeton University Press, Princeton.

Esteva, G., & Prakash, M. S., (1998). Beyond development, what? *Development in Practice, 8*(3), 280–296.

Farmaki, A., (2017). The tourism and peace nexus. *Tourism Management, 59,* 528–540.

Freire-Medeiros, B., (2015). *Touring Poverty.* Abingdon, Routledge.

Gardner, M., (2002). *Harry Truman and Civil Rights.* Southern Illinois University Press, Carbondale.

Gössling, S., & Hultman, J., (2006). *Ecotourism in Scandinavia: Lessons in Theory and Practice* (Vol. 4). CABI, Wallingford.

Hall, C. M., (2011). Policy learning and policy failure in sustainable tourism governance: From first-and second-order to third-order change? *Journal of Sustainable Tourism, 19*(4, 5), 649–671.

Hultman, J., & Hall, C. M., (2012). Tourism place-making: Governance of locality in Sweden. *Annals of Tourism Research, 39*(2), 547–570.

Li, Y., (2000). Geographical consciousness and tourism experience. *Annals of Tourism Research, 27*(4), 863–883.

Litvin, S. W., (1998). Tourism: The world's peace industry? *Journal of Travel Research, 37*(1), 63–66.

Liu, A., & Pratt, S., (2017). Tourism's vulnerability and resilience to terrorism. *Tourism Management, 60,* 404–417.

Mansfeld, Y., & Pizam, A., (2006). *Tourism, Security and Safety.* Routledge, London.

McMichael, P., (2011). *Development and Social Change: A Global Perspective: A Global Perspective.* Sage Publications, Thousand Oaks.

Milano, C., Novelli, M., & Cheer, J. M., (2019). *Over Tourism and Tourism Phobia: A Journey Through Four Decades of Tourism Development, Planning and Local Concerns.* CABI, Wallingford.

Moscardo, G., (2011). The role of knowledge in good governance for tourism. In: Laws, E., Richins, F., & Agrusa, F., (eds.), *Tourist Destination Governance: Practice, Theory and Issues* (pp. 67–80). CABI, Wallingford.

Pizam, A., & Fleischer, A., (2002). Severity versus frequency of acts of terrorism: Which has a larger impact on tourism demand? *Journal of Travel Research, 40*(3), 337–339.

Scheyvens, R., (1999). Ecotourism and the empowerment of local communities. *Tourism Management, 20*(2), 245–249.

Skoll, G. R., (2016). *Globalization of American Fear Culture: The Empire in the Twenty-First Century.* Palgrave Macmillan, Basingstoke.

Sofield, T. H., (2003). *Empowerment for Sustainable Tourism Development.* Emerald Group Publishing, Wagon Lane.

Sönmez, S. F., (1998). Tourism, terrorism, and political instability. *Annals of Tourism Research, 25*(2), 416–456.

Sugiyarto, G., Blake, A., & Sinclair, M. T., (2003). Tourism and globalization: Economic impact in Indonesia. *Annals of Tourism Research, 30*(3), 683–701.

Tarlow, P. E., (2006). A social theory of terrorism and tourism. *Tourism, Security and Safety from Theory to Practice* (pp. 33–48). Elsevier, Bourlington.

Tarlow, P., (2014). *Tourism Security: Strategies for Effectively Managing Travel Risk and Safety.* Elsevier, Oxford.

Tzanelli, R., (2013a). *Heritage in the Digital Era: Cinematic Tourism and the Activist Cause.* Routledge, Abingdon.

Tzanelli, R., (2013b). *Olympic Ceremonialism and the Performance of National Character: From London 2012 to Rio 2016.* Palgrave Macmillan, New York.

Tzanelli, R., (2016). *Thana Tourism and the Cinematic Representation of Risk.* Routledge, Abingdon.

Volgger, M., & Pechlaner, H., (2014). Requirements for destination management organizations in destination governance: Understanding DMO success. *Tourism Management, 41*, 64–75.

Conclusion

A book about the future of something is an arduous task to write. Not only are we having serious disputes revolving around the epistemological origin of the term, but we also have not reached a consensus to define what tourism is. As discussed in several parts of this editorial project, this book is not an attack on any colleague but is a desperate attempt to shed light on the academic community in times of crisis and uncertainty.

Contemporary research is oriented to what some critical voices dubbed an economic-centered paradigm, which means an excess of enthusiasm to the promises of the market and the industry. For this position, tourism is not only a vehicle towards durable peace but also enhances the political stability of democracy. In fact, for the economic-based theory, tourism and democracy are inextricably intertwined. These studies have historically focused on quantitative methods that pose the voice of the Tourist as the only valid source of information. In these investigations, open or closed-ended questionnaires are applied to tourists who are consulted at bus stations, airports, or hotels. Of course, tourists are an essential element of the system but not the only one. As we have argued, sometimes interviewees are incognizant of their inner feelings or simply cover their sentiments to protect their interests.

Over recent years, some scholars have called to adopt qualitative methods and other views coming from sociology and anthropology. To some extent, sociology has developed a pejorative view of tourism, a position that comes from the founding parents of the discipline. In Durkheim (1897, 1972), Weber (2009), or Marx (2000), leisure occupied a little position in their development. The sphere of leisure never captivated the attention of classic sociology because it was esteemed as an objectifying mechanism of control. Sociologists, instead, coordinated efforts to study the role of anomie and social ties as the pre-configuration of society. All of them devoted considerable efforts to understand or answer the question: how is a society working together? The invisible social tie gave them some hints on how solidarity and reciprocity were articulated to form political authority. Until the works of Thorstein Veblen (2017)., Erich

Weber (1969), or Norbert Elias, and Erich Dunning (2018), sociologists scornfully relegated leisure as a bit-player in the formation of social fact.

Having said this, we posit that no less true were modern sociologists who turned their attention to tourism that adopted this critical eye to define tourism. Based on the works of historians of tourism who traced back the origin of the industry to modernity, they overlooked ancient forms of non-Western forms of tourism as a pastime. Once the Roman Empire collapsed, the Middle Age emerged as a dark epoch that lacked mobility and travel. The act of travel was dangerous and often avoided by farmers. However, a closer look suggests the opposite. Romans, Babylonians, and Assyrians constructed complex nets of paths to support a mobile infrastructure to move their armies in the war and also allowing the flux of travelers in peacetime. To some extent, warfare and mobility were inextricably intertwined. The term *Ferias* (Lat.) was used to give three months of leave to all Roman citizens to visit relatives and friends in the Roman provinces. Today, in modern languages such as German or Portuguese, ferias that mutate to Die Ferien (Ger.) or Das Ferias (Por.) merely means holidays. Anthropology, instead, provided some vital knowledge to denote tribal organizations' travels for escapement or relax purposes. This begs the pungent question to what extent tourism is limited to modernity or hypermodernity.

This book defines tourism as a rite of passage that is inscribed into the leisure sphere. Like dreams that revitalize in the nights our frustrations inverting our reality, tourism evolves according to the figure of play and escape. For that reason, the first, second, and third chapters of this book are reserved for debating the nature and evolution of leisure, the contributions and limitations of those theorists who defined tourism earlier than this book, and an explanation that clarifies how rites of passage work. Far from being a ritual performed by tribal societies, anthropology widely showed how rites of passage are present in modern communities. Tourism seems to be understood as a liminal space that not only transforms the personal status but invites the candidate to live in an imagined landscape where he or she emulates to be what, in reality, he or she is not.

The fourth chapter explores the problems and benefits of ethnography as the main method of research or at least to overcome the obstacles that the economic-based paradigm failed to overcome. What do you think a gangster will respond when you ask, what is your profession? He will say, businessman. This comes from a vicious inherited from positivism that is assumed only by asking can we reach information. The ethnography—as a method of inquiry—is more than only asking; it entails seeing, hearing, and writing. However, ethnography has its limitations, which are adamantly debated in the chapter.

The fifth chapter critically reviews the role of conflict and post-conflict destinations. Over the years, researchers applauded the idea that tourism was the precondition to the economic prosperity or development of downgraded economies. Based on the development theory, which advocate that rich countries should morally help other countries, tourism acted as a fertile ground towards fairer wealth distribution. Recently, some scholars have questioned this thesis, arguing convincingly under some conditions that tourism is promoting some long-dormant conflicts, unless regulated, and may very well lead to war.

Doubtless, a book regarding tourism cannot be complete if the scourge of terrorism is not addressed. The sixth chapter deciphers the complexity of terrorism, as well as political violence, to clarify why international tourists are being targeted by radicalized groups. To some extent, 9/11 and terrorism have changed the tourism industry forever. We describe how the literature has experienced changes according to three foundational events: Luxor Massacre, 9/11, and November 2015 Paris attacks. Our thesis punctuates that terrorism not only created millionaire losses in the industry but also modified the Western hospitality closing nations and affected the ways the Non-Western Other is imagined and interpreted. To our end, the anti-tourist sentiment, associated with the restrictive measures adopted by the US to restrict migration or simply Islamophobia, is part of the same phenomenon: the end of hospitality as we know it. This anti-tourist sentiment can be expressly epitomized in a prominent position of artificial intelligence (AI), robots, and humanoids occupied in the tourism research. This point, which is examined in the seventh chapter, not only continues the thesis that *the sacred law of hospitality* – as it was imagined in ancient times – is in jeopardy, but also it mutated to a hybrid version we dubbed as *failed hospitality*, where the (non-human) "Other" is neglected, fabricated, and subordinated to the tourist's pleasure maximization.

We analyze the film plot of *The Passengers,* originally starred by Jennifer Lawrence, Chris Patt, Michael Sheen, and Laurence Fishburne. In a nutshell, this film reveals how in these hybrid spaces, human nature is replaced by a machine while impeding genuine hospitality.

The recent outbreak of COVID-19, a strange virus that spread in China at the end of 2019, is wreaking havoc in the global economies, even in service sectors and the tourism industry. The last chapter speaks to readers of the tourism industry in crisis; there, the world simply stopped. We question here the idea of whether this is the end of tourism and hospitality or simply the rise of new forms of tourism based on virtual technology. In dialog with the former chapter, this section calls attention to the anthropological role of a radical quarantine to struggle against COVID-19. Science has declared war on Coronavirus disease, asking for a global quarantine as the temporal solution. Like in wars where tourism stops and mutates to help the army, hotels are being recycled to be hospitals to contain aggravated clinical pictures. Anthropologically speaking, the lesson of this virus seems to be that the commercial (or failed) hospitality that was demonized by classic sociology sets the pace to a new opportunity towards unconditional hospitality. As a social phenomenon, tourism never disappears; rather, it goes towards new directions according to the transformation of mainstream cultural values of society. Hopefully, the present academic work clarifies scholars and academia in a moment of chaos and crisis.

REFERENCES

Durkheim, E. (1897). *Le suicide: étude de sociologie*. Alcan, Paris.

Durkheim, E., (1972). *Emile Durkheim: Selected Writings*. Cambridge University Press, Cambridge

Elias, N., & Dunning, E. (2008). *Quest for excitement: Sport and leisure in the civilising process*. University College Dublin Press, Dublin.

Marx, K. (2000). *Karl Marx: Selected Writings*. Oxford University Press, Oxford.

Veblen, T. (2017). *The Theory of the Leisure Class*. Routledge, Abingdon.

Weber, M. (2009). *From Max Weber: Essays in Sociology*. Routledge, Abingdon.

Weber, E (1969). *El problema del tiempo libre (The Problems of Leisure)*. Editorial Nacional, Madrid.

Index

For Product Safety Concerns and Information please contact our EU
representative GPSR@taylorandfrancis.com
Taylor & Francis Verlag GmbH, Kaufingerstraße 24, 80331 München, Germany

www.ingramcontent.com/pod-product-compliance
Lightning Source LLC
Chambersburg PA
CBHW060349220326
41598CB00023B/2858